AF389427

Smart Management of Conservative, Organic and Integrated Agriculture

Smart Management of Conservative, Organic and Integrated Agriculture

Editors

Andrea Peruzzi
Christian Frasconi
Daniele Antichi

MDPI • Basel • Beijing • Wuhan • Barcelona • Belgrade • Manchester • Tokyo • Cluj • Tianjin

Editors
Andrea Peruzzi
University of Pisa
Italy

Christian Frasconi
University of Pisa
Italy

Daniele Antichi
University of Pisa
Italy

Editorial Office
MDPI
St. Alban-Anlage 66
4052 Basel, Switzerland

This is a reprint of articles from the Special Issue published online in the open access journal *Agronomy* (ISSN 2073-4395) (available at: https://www.mdpi.com/journal/agronomy/special_issues/smart_management_conservative_agriculture).

For citation purposes, cite each article independently as indicated on the article page online and as indicated below:

LastName, A.A.; LastName, B.B.; LastName, C.C. Article Title. *Journal Name* **Year**, *Article Number*, Page Range.

ISBN 978-3-03936-986-7 (Hbk)
ISBN 978-3-03936-987-4 (PDF)

Cover image courtesy of Andrea Peruzzi.

Contents

About the Editors . . . . **vii**

Preface to "Smart Management of Conservative, Organic and Integrated Agriculture" . . . . . **ix**

Ana I. Marí, Gabriel Pardo, Alicia Cirujeda and Yolanda Martínez
Economic Evaluation of Biodegradable Plastic Films and Paper Mulches Used in Open-Air
Grown Pepper (*Capsicum annum* L.) Crop
Reprinted from: *Agronomy* **2019**, *9*, 36, doi:10.3390/agronomy9010036 . . . . **1**

**Luisa Martelloni, Michele Raffaelli, Christian Frasconi, Marco Fontanelli, Andrea Peruzzi
and Claudio D'Onofrio**
Using Flaming as an Alternative Method to Vine Suckering
Reprinted from: *Agronomy* **2019**, *9*, 147, doi:10.3390/agronomy9030147 . . . . **15**

Laura Vincent-Caboud, Léa Vereecke, Erin Silva and Joséphine Peigné
Cover Crop Effectiveness Varies in Cover Crop-Based Rotational Tillage Organic Soybean
Systems Depending on Species and Environment
Reprinted from: *Agronomy* **2019**, *9*, 319, doi:10.3390/agronomy9060319 . . . . **29**

Simona Bosco, Iride Volpi, Daniele Antichi, Giorgio Ragaglini and Christian Frasconi
Greenhouse Gas Emissions from Soil Cultivated with Vegetables in Crop Rotation under
Integrated, Organic and Organic Conservation Management in a Mediterranean Environment
Reprinted from: *Agronomy* **2019**, *9*, 446, doi:10.3390/agronomy9080446 . . . . **47**

**Lara Abou Chehade, Daniele Antichi, Luisa Martelloni, Christian Frasconi, Massimo Sbrana,
Marco Mazzoncini and Andrea Peruzzi**
Evaluation of the Agronomic Performance of Organic Processing Tomato as Affected by
Different Cover Crop Residues Management
Reprinted from: *Agronomy* **2019**, *9*, 504, doi:10.3390/agronomy9090504 . . . . **73**

**Robson da Costa Leite, José Geraldo Donizetti dos Santos, Rubson da Costa Leite,
Luciano Fernandes Sousa, Guilherme Octávio de Sousa Soares, Luan Fernandes Rodrigues,
Jefferson Santana da Silva Carneiro and Antonio Clementino dos Santos**
Leguminous Alley Cropping Improves the Production, Nutrition, and Yield of Forage Sorghum
Reprinted from: *Agronomy* **2019**, *9*, 636, doi:10.3390/agronomy9100636 . . . . **87**

Ted S. Kornecki and Andrew J. Price
Management of High-Residue Cover Crops in a Conservation Tillage Organic Vegetable
On-Farm Setting
Reprinted from: *Agronomy* **2019**, *9*, 640, doi:10.3390/agronomy9100640 . . . . **103**

**Giacomo Tosti, Paolo Benincasa, Michela Farneselli, Marcello Guiducci, Andrea Onofri and
Francesco Tei**
Processing Tomato–Durum Wheat Rotation under Integrated, Organic and Mulch-Based
No-Tillage Organic Systems: Yield, N Balance and N Loss
Reprinted from: *Agronomy* **2019**, *9*, 718, doi:10.3390/agronomy9110718 . . . . **117**

**Daniele Antichi, Massimo Sbrana, Luisa Martelloni, Lara Abou Chehade, Marco Fontanelli,
Michele Raffaelli, Marco Mazzoncini, Andrea Peruzzi and Christian Frasconi**
Agronomic Performances of Organic Field Vegetables Managed with Conservation Agriculture
Techniques: A Study from Central Italy
Reprinted from: *Agronomy* **2019**, *9*, 810, doi:10.3390/agronomy9120810 . . . . **129**

Aldo Calcante and Roberto Oberti
A Technical-Economic Comparison between Conventional Tillage and Conservative Techniques
in Paddy-Rice Production Practice in Northern Italy
Reprinted from: *Agronomy* **2019**, *9*, 886, doi:10.3390/agronomy9120886 . . . . **157**

About the Editors

Andrea Peruzzi graduated in Agricultural Sciences (MS.c.) summa cum laude in 1983, and in 1989, completed his Ph.D. in Agricultural Engineering;

- he is a member of EurAgEng and CGIR, EWRS, ISTRO and "Accademia dei Georgofili";
- actually he is Full Professor of "Applied Physics", "Agricultural Engineering and Farm Mechanization" and "Mechanization in organic farming" at the University of Pisa;
- from 1995, he has been the visiting professor in several foreign research institutions;
- he has been, and is, the scientific responsible and/or coordinator of many national and international research projects;
- the main research subjects concern the definition of strategies and the design and full realization of innovative machines for conservation tillage and direct planting, non-chemical weed control and crop protection in organic farming, according to a sustainable management of agriculture. The research work is supplied by more than 500 scientific papers.

Christian Frasconi Ph.D., graduated in Agriculture (MS.c.) at the University of Pisa in 2006, and received the title of Doctor of Philosophy at the University of Florence under the Ph.D. in Agriculture and Forestry Engineering in 2010. Since October 2018, he has been an assistant professor at the Department of Agriculture, Food and Environment of the University of Pisa.
He is member of:
Associazione Italiana di Ingegneria Agraria (AIIA);
Commission Internationale du Genie Rural (CIGR);
European Society of Agricultural Engineers (EurAgEng);
His teaching activities are related to farm machinery and farm mechanization, applied physics and mechanization in organic farming.
His main research subjects concern innovative machines and strategies for non-chemical weed control, conservation agriculture and cover crop management.

Daniele Antichi Ph.D., graduated in Agriculture (MS.c.) at the University of Pisa in 2003, and received the title of Doctor of Philosophy at the University of Pisa under the Ph.D. in Science of Crop Production Cycle XXIV. Since January 2016, he has been an assistant professor at the Department of Agriculture, Food and Environment of the University of Pisa;
- he is a member of EWRS, SIA, RIRAB, and Agroecology Europe;
- he is Principal Investigator for the University of Pisa of the EU Horizon—2020 Projects, IWMPRAISE, LegValue, AGROMIX;
- his teaching activities are related to organic arable crop production (MS.c. at the University of Pisa);
- his main research subjects concern sustainable farming, conservation agriculture, organic farming field crop production, cover cropping and intercropping.

Preface to "Smart Management of Conservative, Organic and Integrated Agriculture"

Sustainable agriculture is targeted towards achieve food security, while maximizing the socio-economic benefits and minimizing environmental drawbacks. Among sustainable farming practices, organic and integrated farming systems are widely recognized as effective farming systems, in terms of global warming mitigation and contrast to soil desertification. As a matter of fact, certified organic land in Europe has increased by almost 75% in the last decade. Globally, the increasing demand for environmental sustainability, safety and food quality surely encourage farmers to change their agricultural strategies moving from "conventional" (i.e., intensive, market-oriented, agro-industrial systems) to integrated and organic farming. The first step of this transaction is a reduction in the use of external chemical inputs (e.g. mineral fertilizers, synthetic pesticides). However, both organic and integrated farming require a complete shift in the agricultural management approach, to fully express their potential. Actually, many farmers converting to organic farming rely on the so-called "Input Substitution Approach", a simplified management approach, based mostly upon replacing synthetic agrochemicals with natural substances allowed by the organic farming regulations. Normally, intensive tillage is also practiced for seed bed preparation, organic fertilizer/green manure/crop residue incorporation and, although to a lesser extent, weed management, thus hindering to achieve one of the key objective of organic farming, i.e., to conserve and improve soil fertility. Intensive tillage can deplete soil organic matter, could be responsible for soil erosion through the destruction of soil structure, and can decrease soil biological activity and biodiversity.

On the other hand, conservation agriculture (CA), defined according to the Food and Agriculture Organization of the United Nations (FAO) as the combination of reduced soil disturbance, permanent soil cover and diversification of cropping systems, is a rising management system reputed to: reduce the risks of erosion and nutrient loss, increase soil organic matter and carbon sink capacity, improve soil fertility and contrast global warming. Reduced and no-tillage systems were developed a few decades ago in conventional agriculture, to pursue these goals, as well as obtain relevant energy and economic savings, by eliminating huge tillage and excessive field traffic. With this aim, many research efforts have been spent to design and realize operative machines able to perform no-tillage in an appropriate and effective way (i.e., no-till drills, planters and trans-planters) and reduced tillage on entire fields or in band (i.e., strip tillage implement). All these machines are equipped with tools suitable to allow a good preparation of the seed-bed and a proper management of soil cover. Unfortunately, CA generally relies on the large-scale use of agrochemicals, with a reduction of energy efficiency and an increase in environmental impact. For these reasons, introducing CA techniques into organic farming could be really challenging if compared to integrated farming systems, where the application of agrochemicals is limited, but still allowed.

However, recent studies demonstrated that the application of CA techniques in organic farming could be facilitated by the use of different typologies of mulch (although this solution often resulted in a high increase of cultivation costs, negatively influencing farmers income) and/or by the inclusion of cover crops in crop rotations. Using legume species as cover crops also improves N nutrition of the cash crop and increase soil nitrogen organic pool. For these reasons, in recent times, researchers increased the investigations on cover-crop-based reduced and no-till farming systems, as a sustainable practice to eliminate the reliance on intensive tillage, and maximize the benefits of cover crops and resource use efficiency in organic farming. In these systems, cover crops are often terminated without incorporating residues into the soil, thus leaving a dead mulch, into which the cash crop is planted using appropriate tailor-made operative machines, able to properly work on reduced or no tilled soil covered by dead mulch. This requires the necessity to produce large cover crop biomass, as well as a good management of their residues to provide maximum weed suppression and nutrient cycling. Weed management and nutrient availability are two factors known to challenge the crops performance in organic and conservative production. As a matter of fact, weed pressure tends to increase, although cover crops can reduce weed infestation during their growth, making a physical barrier consisting of dead mulch on the soil surface, preventing sunlight reaching the soil surface and through allelopathy. However, the important results obtained in many recent researches on the set up of strategies and the design and realization of machines for physical weed control will surely allow one to define valid solutions in all agricultural contests, to solve this "key problem".

In conclusion, in this Special Issue as Guests Editors we decided to take into consideration all the researches concerning with the definition and testing of smart solutions, based on the use of both agronomic strategies and innovative agricultural machinery, related to the proper management of organic, integrated and conservation farming systems, taken both alone and together. We are really satisfied with the final results, as the papers published in this SI surely added relevant and innovative knowledge for the smart management of organic and conservative agriculture.

However, going into detail, the 10 papers published in this SI concern research on:

- smart management of farming systems, based on combination between conservation agricultural practices and organic management of vegetable and arable crops, with the inclusion of cover crops and appropriate strategies and machines for their termination and for weed control (6),

- only strategies to be used in organic farming: use of plastic and paper mulches to control weed in pepper, use of in-row flaming for weed and sucker control in the vineyard, use of leguminous alley cropping in sorghum (3),

- only conservation tillage practices: a technical-economic comparison between conservation and conventional tillage in paddy-rice (1).

Andrea Peruzzi, Christian Frasconi, Daniele Antichi
Editors

Article

Economic Evaluation of Biodegradable Plastic Films and Paper Mulches Used in Open-Air Grown Pepper (*Capsicum annum* L.) Crop

Ana I. Marí [1,*], Gabriel Pardo [2], Alicia Cirujeda [2] and Yolanda Martínez [3]

[1] Department of Plant Health, Weed Laboratory, Centro de Investigación y Tecnología Agroalimentaria de Aragón (CITA), Avda. Montañana 930, ES 50059 Zaragoza, Spain

[2] Department of Plant Health, Weed Laboratory, Centro de Investigación y Tecnología Agroalimentaria de Aragón-IA2 (CITA-University of Zaragoza), Avda. Montañana 930, ES 50059 Zaragoza, Spain; gpardos@aragon.es (G.P.); acirujeda@aragon.es (A.C.)

[3] Department of Economic Analysis, Centro de Investigación y Tecnología Agroalimentaria de Aragón-Instituto Agroalimentario de Aragón-IA2 (University of Zaragoza-CITA), Gran Vía 2-4, 50004 Zaragoza, Spain; yolandam@unizar.es

* Correspondence: aimari@aragon.es; Tel.: +34-976-71-41-01

Received: 17 December 2018; Accepted: 14 January 2019; Published: 16 January 2019

Abstract: Black polyethylene (PE) is the most common mulching material used in horticultural crops in the world but its use represents a very serious environmental problem. Biodegradable films and paper mulches are available alternatives but farmers are reluctant to adopt them because of their high market prices. The aim of this paper is to evaluate the economic profitability of eight biodegradable mulching materials available for open-air pepper production. The economic evaluation is based on a four-year trial located in a semi-arid region of Spain. Three scenarios of PE waste management are examined: (i) absence of residues management, (ii) landfill accumulation, and (iii) total recycling. The inclusion of the costs of waste management and recycling under the current Spanish legislation only reduced the final net margin by 0.2%. The results show that an increase in subsidy rates of up to 50.1% on the market price would allow all biodegradable films to be economic alternatives to PE. The study supports the mandatory measures for the farmers to assume the costs of waste management and recycling. Despite savings in field conditioning costs, high market prices of biodegradable materials and papers are not compensated by the current level of subsidies, hampering their adoption in the fields.

Keywords: waste management; economic evaluation; biodegradable mulch; polyethylene

---

1. Introduction

Mulching materials have demonstrated many advantages in controlling weeds, [1,2] increasing soil temperature [3] and moisture [4] and reducing soil degradation [3]. These features finally influence in increasing crop yields [5]. In general, the literature recognizes that all these effects have positive outcomes on economic profitability because of water savings (up to 25%) and reduced labor costs for weed and pest control. [6–8]

Despite all these reported advantages, two major problems threaten such savings at a short and long-term. First, mulch application, removal, and disposal are labor-intensive and hence costly [9,10], and second, the most commonly used mulching materials (polyethylene and other fuel-based films) involve environmental risks in the long-term because their chemical structure is difficult to degrade [11]. The negative environmental effects [12] include the persistence of unrecovered plastic mulch in soil, their potential to alter soil quality by accelerating carbon and nitrogen metabolism, as well as potentially degrading soil organic matter. The presence of plastic residues in the soil can cause significant losses in production. For example, [13] reported that plant growth and yield of tomato crop were affected significantly when residual plastic mulch in soils reaches 160 kg ha^{-1}.

The most frequently used mulching materials in agriculture are manufactured mainly from petrol-based sheets like PE [14], low-density polyethylene (LD-PE) and linear low-density polyethylene (LLD-PE). These types of materials account for 17.5% of total demand by resin types in Europe [15]. The main tool to control weeds in vegetable crops is LD-PE film because it is a very cheap and easy-to-use material [16]. High amounts of waste generated by PE mulches both in the field and in landfills raise many concerns. Although plastic recycling is well established in central Europe, in other countries like Spain, agricultural plastic wastes generate 75,000 tons per year and most of them are tilled into the field, burned, or just left behind in adjacent areas [17–19]. In countries like China [18], it has been reported that the amount of waste in a common vegetable farming field could reach between 50 and 260 kg ha^{-1}. In this context, biodegradable variants of mulching are promising alternatives in vegetable production. The use of such mulches adds to the above-mentioned benefits and additionally reduces disposal costs for farmers while preventing environmental problems in the long-term. These mulching supplies include paper (cellulosic fiber), polylactic acid, polyester and corn, sugar cane, or potato starch [20].

Biodegradable films and paper mulches have been studied previously, demonstrating that productions are statistically the same than obtained with PE [1,21–24]. However, their market prices are higher than PE thus reducing its economic attractiveness for farmers in the short-term. In addition, there are no exhaustive studies including economic evaluations of PE and biodegradable mulches containing (i) an estimation of plastic removal costs; and (ii) a global consideration of short and long-term advantages and limitations of mulching materials [12].

The aim of this paper is to contribute new data to the literature by comparing the economic outcomes of PE and eight different mulching materials available for open-air pepper production. The economic evaluation is based on a four-year trial located in Aragon (Spain) with semi-arid climate conditions. Spain is currently the fifth highest world producer in pepper and the first in Europe [25] with more than 1.1 million annual tons and one of the highest average productivities in the world (6.11 kg m^{-2}). Fresh pepper is the main greenhouse vegetable cultivated in Spain, although the open-air cultivation is widespread in the country.

In order to promote the use of biodegradable materials, some regional authorities in Spain, like the Aragon Government, have implemented economic incentives for farmers who employ biodegradable mulching in vegetable production subjected to some other additional conditions. This study includes these incentives in economic calculations and evaluates their effectiveness in promoting the use of biodegradable mulches. The analysis contributes to the literature by providing data for discussion on the short- and long-term effects of the use of mulching materials.

2. Materials and Methods

2.1. Field Trials and Experimental Design

Field trials were conducted in an experimental field located in Zaragoza, Spain (41.43° N, 0.48° W) from May to October in 2012 to 2015, on a soil with a loamy texture (37.75% sand, 49.08% silt and 13.1% clay), with 2.1% organic matter and pH 7.95. Table 1 shows the main weather parameters during the cropping season in the years of trials.

Table 1. Average monthly temperature (°C), monthly solar radiation (h), solar radiation (MJ m^{-1}), rainfall (mm), days of rainfall, and number of days with gusts >10 m s^{-1} from May to October from 2012 to 2015.

Year	Month	Average Monthly Temperature (°C)	Monthly Solar Insolation (h)	Solar Radiation (MJ m^{-1})	Rainfall (mm)	Days of Rainfall	Number of Days with Gusts >10 m s^{-1}
2012	May	19.8	306	360	3.3	6	4
2012	Jun	23.2	374	443	36.9	6	6
2012	Jul	23.7	395	467	2.8	3	5
2012	Aug	25.7	363	389	0.1	1	5
2012	Sep	20.3	305	252	18.5	6	7
2012	Oct [i]	17.0	164	97	12.6	3	3
2013	May	13.7	253	708.78	29	12	10
2013	Jun	19.6	285	769.9	32.9	5	8
2013	Jul	25.5	335	824.7	35.8	12	6
2013	Aug	23.7	312	749.1	17.8	3	3
2013	Sep	20.4	276	567.39	14.1	4	5
2013	Oct	16.9	261	405.82	17.1	7	4
2014	May	16.6	276	773.52	27.05	8	5
2014	Jun	22.0	296	798.61	18.82	8	9
2014	Jul	23.0	334	821.31	0.4	3	9
2014	Aug	23.2	308	739.53	12.06	5	5
2014	Sep	21.6	258	531.14	23.02	8	3
2014	Oct	17.3	250	388.8	9.02	6	3
2015	May	18.5	380.5	781.7	3.93	4	11
2015	Jun	22.7	371	808.01	24.31	8	8
2015	Jul	25.9	380.6	785.5	13.13	4	10
2015	Aug	23.8	355.5	727.14	26.27	10	2
2015	Sep	18.7	310.5	253.9	24.1	6	-
2015	Oct	15.0	260.5	145.7	36.6	14	-
Av.	May	17.2	263	736 *	44	7.5 *	7.5 *
Av.	Jun	21.3	295	797 *	31	6.8 *	7.75 *
Av.	Jul	24.5	337	829 *	18	5.5 *	7.5 *
Av.	Aug	24.4	311	746 *	17	4.8 *	3.75 *
Av.	Sep	20.7	231	475 *	27	6 *	5 *
Av.	Oct	15.5	192	299 *	30	7.5 *	3.3 *

[i] Average only with 18 days; Av. average period 1970–2010; * only average period 2012–2015.

Treatments were distributed randomly in a complete block design with four replicates. Elementary plots measured 0.7 m wide raised beds spaced 1.5 m from center to center and of 20 m longitude. Eight mulches (four biodegradable plastics and four papers) were tested and black polyethylene (PE) plastic was added as a control (Table 2). These materials were selected because they are available on the market, are still in the experimental phase, or have recently been marketed. All materials measured 1.2 m wide and were mechanically installed within five days after soil preparation prior to weed emergence. Soil preparation included soil tillage and bed formation. The irrigation system used was a 16 mm diameter drip tape in each line with an emitter every 20 cm and treatments were grouped into two different sectors, i.e., paper and plastic mulches, which were irrigated separately according to their water needs [26]. The irrigation moment was calculated with the soil moisture sensors (Aquameter ECH2O. Decagon Devices, Washington, DC, USA) thus the plants were irrigated before the stress of the crop (minimum balance) begins. The pepper variety was "Viriato" type Lamuyo. Pepper was transplanted with 0.3 m plant spacing, double row distribution, and 0.3 m between rows of crop. Marketable pepper fruits were harvested three times at the end of the season (during one month in all years).

Data on yield, inputs, and operational costs were collected each year from the trials in order to analyze the economic outcomes of each material. The analysis of yield data was performed using SAS (Statistical Analysis System V.9.4. SAS Institute, Cary, NC, USA). Homogeneity of variance and normality was tested before data analysis. Data were subjected to analysis of variance (ANOVA). Given that p value of ANOVA was higher than 0.05 ($p = 0.45$) mean separations were not performed.

For the economic part of the analysis, the operational costs, incomes, and net margins are presented separately.

Table 2. Type, name, main composition, thickness (μm) (plastic films) or grammage (g m^{-2}) (paper mulches), and color of materials used in the trials.

Type of Mulching	Mulching Materials	Main Composition	Thickness–Grammage (μm–g m^{-2})	Color
Non-degradable plastic film	PE	Low-density polyethylene	15	Black
Biodegradable films	Mater-Bi[®1]	Polycaprolactone, starch blend	15	Black
	Sphere[®2]	Potato starch, recycled polymers	15	Black
	Bioflex[®3]	Polylactic acid, co-polyester	15	Black
	Ecovio[®4]	Polylactic acid, polybutylene adipate terephthalate, starch	15	Black
Paper	Arrosi[®] 69[5]	Cellulosic fiber	80	Light brown
	Arrosi[®] G1a[5]	Cellulosic fiber	100	Light brown
	Arrosi[®] 240[5]	Cellulosic fiber	80	Light brown
	Mimgreen[®6]	Cellulosic fiber	85	Black

[1] Novamont S.p.A. Novara, Italy. [2] Sphere Group Spain S.L. Zaragoza, Spain. [3] FKuR Kunststoff GmbH. Willich, Germany. [4] Fábrica de Papeles Crepados Arrosi S.A. Gipuzkoa, Spain. [6] Mimcord S.A. Barcelona, Spain.

2.2. Costs

Table 3 shows the inputs used and operational costs considered including fuel consumption. Inputs costs include pepper seedlings, pre-transplanting manure, herbicides, chemical dressing, irrigation water, and mulching materials used in trials. Pre-transplanting manure, chemical dressing, and some field preparation labors were taken from the experimental trial and the rest of the time costs considered for each operation were obtained from an interview with a local pepper producer. Labor costs are calculated using official data available in [27]. Amounts and type of fertilizers and doses of active matters used in chemical dressing can be consulted in [28].

Prices of mulching materials were obtained directly from the manufacturers thus they are final market prices. The costs of mechanical installation of paper mulches were calculated using data published by [1] for the case of tomato crop, adding an extra cost derived from the considered speed in the specific case of paper mulches, which need to be installed slower because they are not flexible and break easily. Additionally, a PE roll usually contains 2400 linear meters while a paper roll contains approximately 250 linear meters. Therefore, the number of times that workers have to stop to change roller in order to mulch a field of the same surface has also been considered. Similarly, the time needed to bury the endpoint of the mulch in each line in order to fix the material to the soil is considered.

Irrigation costs include an annual quota (proportional to the amount of hectares), energy costs, and drip line purchase cost. Operational costs include labor and machinery costs for soil preparation, crop and mulching installation and removal, application of fertilizers and herbicides, harvesting, and final field conditioning.

The cost of transplanting operation varies depending on the hired company and its availability at the time of the operation. Hence, an average costs from two different local companies was used. Chemical dressing was applied by fertirrigation and fractioned 6 times and labor cost was included. Herbicide application between line crops and manual weeding in the transplanting holes are common tasks and the costs are quite variable among years so an average rate provided by the farmer was used. Harvesting is one of the most expensive operations in the case of pepper for fresh consumption because the fruits are manually collected between three to four times at the end of the cropping season.

Table 3. Costs ($€\ ha^{-1}$) of inputs and operations in open-air pepper production.

Inputs		Cost ($€\ ha^{-1}$)
Pepper seedlings		1350
Pre-transplanting manure		900
Herbicides		24.3
Chemical dressing		810
Irrigation	Annual payment	123
	Electric consumption	290
	Drip line	238
Mulches [a]	PE	404
	Mater-Bi®	1164
	Sphere®	772
	Bioflex®	931
	Ecovio®	505
	Mimgreen®	1086
	Arrosi® 69	1024
	Arrosi® G 1a	1358
	Arrosi® 240	1024
Operations		
Subsoiler		113
Cultivator tillage		51
Rotatory tiller		230
Pre-transplanting manure application		103
Burying fertilizer		51
Installation irrigation system		244
Bed formation + drip line installation + plastic mulching		144
Bed formation + drip line installation + paper mulching		178
Crop installation/transplant		475
Chemical dressing application		17.5
Herbicide between lines		9
Manual weeding transplanting holes		350
Manual harvest		2340
Irrigation system removal		130
Crop removal		51
PE removal		176.5
Landfill [b]		186
Recycling [b]		192
Cultivator tillage		51

[a] For 0.7 m bed width and 1.5 separation between lines; [b] For a plastic consumption of 160 kg ha^{-1}; Management of plastic, transport time, landfill and recycling costs included.

Field conditioning involves manual removal of the irrigation system, crop rests removal (which is a combined mechanical and manual operation) and plastic elimination in the case of non-biodegradable films which is a mechanical operation with a rotatory machine coupled to the tractor. The cost of landfill must be considered because under the current Spanish Law, farmers are responsible of ensuring proper treatment of wastes produced in their fields. However, as they are not required to assume the cost of recycling farmers usually store their waste and transport it to an authorized recovery point. Although recycling is not mandatory for farmers in Spain, we consider a scenario of plastic recycling in order to evaluate its effect on the final profitability. As a consequence, three different scenarios are considered: (i) the most widespread situation where farmers do not conduct any waste treatment, just remove the plastic residues from the field and leave them stored, buried or burned; (ii) the landfill scenario, where farmers transport plastic residues to the recovery point, and (iii) the recycling situation, when the farmers transport the residues to the recycling plant and assume the recycling cost. The consideration of the no waste treatment as a baseline scenario will allow us to assess how profitability is affected by waste treatment, which is a contribution of this paper.

The costs of manipulation and transport (including fuel) of the plastic waste from field to the recovery point (or the recycling plant) are included in scenarios (ii) and (iii) as an externalized task. This cost includes plastic removal from the field with a specific rotatory machine and the transport of the residues to the final destination with a tractor provided with a tow. A distance of 30 km from the field to the recovery point has been considered for the calculations. For the recycling scenario, the cost was obtained from a local recycling plant which amounts 62 € t^{-1}. Usually, film mulches have impurities such as soil, debris, pesticides, or fertilizers, which can represent up to 85% of the total remnants by weight and recycling plants usually do not accept plastic films with more than 5% impurities [29]. However, the local plant considered does not establish a limit for impurities.

Finally, cultivator tillage cost for soil preparation for the next season is included as field conditioning. Costs of using machinery shown in Table 1 includes the cost of fuel which is proportionally distributed in proportion to the time cost of each operation.

2.3. Incomes and Net Margins

The calculation of incomes includes the market value for the crop outputs. The "Lamuyo" pepper market price considered is 876 € t^{-1}, which is an average from the last three years from available data [27]. We assume that this market price is not different between materials because we have not observed that different mulches modifies the harvest time in the case of pepper crop.

Although there were no statistical differences among materials [28], yields obtained in three to four years of the experiment were very low (about 10 t ha^{-1}) in comparison to the average obtained in the region which amounts 29.8 t ha^{-1} [30]. Pepper is a delicate crop concerning water and humidity variations and during 2012 and 2013, technical problems in irrigation caused pepper seedlings mortality that could not be replaced. In addition, 12 days of rainfall were reported in 2013 (7.5 days is the usual) (see Table 1). Although the amount of rainfall was not excessive, it caused a delay in the field works, which led to planting peppers to a very late date (15 June). This is a handicap to get good production in our area.

In 2015, temperature, insolation, and radiation parameters during May and June were much higher than normal, which caused the degradation of many biodegradable plastics and thinner papers and interfering dramatically with flowering. Subsequently these materials broke more easily by the action of the wind, which was also stronger than usual from May to October if we look at the days of wind with gusts greater than 10 m s^{-1}.

Therefore, yield data used in this study is from year 2014 where pepper yields are considered normal compared to the average production in the area and no agronomic and climatic problems were observed.

Additionally, farmers can obtain subsidies from the Aragon Government (funded by the European Union) offering the possibility to receive 35% of the material costs when biodegradable mulching is used. In such case, farmers must also meet some demanding requirements, such as belonging to a horticultural producers' association developing operative and investment programs in improving the quality of their products including the development of protected designations of origin and geographical indications [31]. According to current legislation, paper mulches are not considered as biodegradable and therefore do not receive subsidies. Consequently, two different scenarios are considered in the economic analysis: (i) when no subsidies are received; (ii) when farmers are compensated for the cost of using biodegradable mulches. This comparison sheds light on practical insights to improve the knowledge of the effectiveness of such subsidies in promoting the use of biodegradable materials.

Finally, the economic profitability of each material is compared using the net margin, which is calculated as the difference between incomes (value of the crop output with or without regional subsidies) and total costs (inputs, operations, labor, etc.).

3. Results

3.1. Costs and Incomes

Comparing the cost of the considered mulches, biodegradable materials are between 25% and 188% more expensive than PE while paper mulches are between 153% and 236% more expensive (see Table 3). Among biodegradable materials, Ecovio® is the cheapest one and Arrosi® 69 and Arrosi® 240 are the cheapest papers.

Table 4 shows the aggregated costs by operations calculated in the trials. The name "field preparation" includes subsoiler, cultivator tillage, rotatory tillering, and the application and burial of pre-transplanting manure. "Crop season operations" comprised irrigation, herbicide application and chemical dressing among others. "Plastic and paper mechanical mulching" includes the costs of materials and mechanical installation on the field. Finally, the concept of "field conditioning" includes irrigation system and crop removal, waste management for the non-biodegradable scenarios, and, finally, a cultivator pass.

Table 4. Costs (€ ha^{-1}) for fresh pepper crop production.

Operations		Costs (€ ha^{-1})
Field preparation		1448
Crop season operations		3931
Plastic mechanical mulching	PE	548
	Mater-Bi®	1308
	Sphere®	916
	Bioflex®	1075
	Ecovio®	649
Paper mechanical mulching	Mimgreen®	1264
	Arrosi® 69	1202
	Arrosi® G 1a	1536
	Arrosi® 240	1202
Harvest		2340
Field conditioning non-biodegradable mulch scenario [a]	No waste management	408.5
	Landfill	418
	Recycling	424
Field conditioning biodegradable mulch scenario		232

[a] For a plastic consumption of 160 kg ha^{-1}. Management of plastic, transport time, and landfill and recycling costs included.

If the use of PE with no waste management is considered as a benchmark, then mulching represents 6.3% of the total costs for pepper production. The biggest expenditure of these operations corresponds to crop season operations (mainly transplant and pepper seedlings costs) with 45.3% and the following is the harvest with 27% because it is a manual task. For the rest of the cases, mulching materials represents between 7.5% and 14.1% of the total costs in biodegradable and between 13.1% and 16.2% in paper types (Table 4). Regarding irrigation costs, although we expected to save water with plastics with respect to papers, water consumption was very similar for both types of materials.

The analysis of field conditioning costs for PE scenario shows that this cost represents 4.7% of the total when no waste management is carried out. This cost increases to 4.8% when the farmer transports the waste to the recycling point (landfill scenario) and up to 4.9% if the complete recycling cost is assumed. By contrast, using biodegradable mulches allows a saving in field conditioning of a minimum of 54.7% and a maximum of 56.7% with respect to PE.

Table 5 shows the results obtained for yield, subsidies, and incomes. Despite no statistically differences are found among mulching materials, PE obtained one of the lowest yields. Mater-Bi® and Arrosi®240 obtained amounts close to 30 t ha^{-1}, which are similar to the average yields recorded in Spain (29 t ha^{-1}). Final incomes were calculated including the subsidies available to cover 35% of the biodegradable plastic cost.

Table 5. Experimental yield (t ha^{-1}), subsidies, and total income obtained for mulching materials in open-air conditions in 2014.

Type of Mulching	Mulching Materials	Yield (t ha^{-1})	Subsidies (€ ha^{-1})	Income with Subsidies (€ ha^{-1})
Non-degradable film	PE	24.6 a	-	21,549.6
Biodegradable films	Mater-Bi$^{®}$	29.2 a	407.4	25,986.6
	Sphere$^{®}$	25.8 a	270.2	22,871.0
	Bioflex$^{®}$	24.4 a	325.9	21,700.3
	Ecovio$^{®}$	23.3 a	176.8	20,587.6
Paper mulch	Mimgreen$^{®}$	26.7 a	-	23,389.2
	Arrosi$^{®}$69	25.3 a	-	22,162.8
	Arrosi$^{®}$G 1a	26.9 a	-	23,564.4
	Arrosi$^{®}$240	28.5 a	-	24,966.0

Same letters in yield mean no statistical differences among treatments ($p = 0.45$).

3.2. Net Margins

Table 6 summarizes the main economic variables analyzed. Net margins are calculated under the three waste management scenarios considered for PE and under the two scenarios for biodegradable materials (with and without subsidies). In addition, the percentage with respect to PE without waste management (baseline scenario) is calculated in order to present a comparative analysis of alternative materials.

For biodegradable materials, the total costs are between 2.2% and 9.3% higher than those of PE. The only exception is Ecovio$^{®}$, which is cheaper than PE because the additional material cost is less than disposal costs. Regarding final profitability, two bio-degradable materials (Mater-Bi$^{®}$ and Sphere$^{®}$) present higher profitability than PE (with and without subsidies) while Bioflex$^{®}$ and Ecovio$^{®}$ are the worse options, with reductions of 1.6% and 6.9% with respect to the benchmark due to low yields obtained in the trials. Mater-Bi$^{®}$ is the best biodegradable option, with an increase of 29.9% with respect to PE.

Table 6. Incomes, costs, and net margins of different mulching materials (€ ha^{-1}).

Type of Mulching	Mulching Materials	Scenarios	Incomes	Costs	Net Margin	% with Respect to PE
Non-degradable film	PE	No waste management	21,549.6	8675.3	12,874.3	-
		Landfill	21,549.6	8684.8	12,864.8	99.9
		Recycling	21,549.6	8690.8	12,858.8	99.9
Biodegradable films	Mater-Bi$^{®}$	No subsidies	25,579.2	9258.8	16,320.4	126.8
		With subsidies	25,986.6		16,727.8	129.9
	Sphere$^{®}$	No subsidies	22,600.8	8866.8	13,734.0	106.7
		With subsidies	22,871.0		14,004.2	108.8
	Bioflex$^{®}$	No subsidies	21,374.4	9025.8	12,348.6	95.9
		With subsidies	21,700.3		12,674.5	98.4
	Ecovio$^{®}$	No subsidies	20,410.8	8599.8	11,811.0	91.7
		With subsidies	20,587.6		11,987.8	93.1
Paper	Mimgreen$^{®}$	No subsidies	23,389.2	9214.8	14,174.4	110.1
	Arrosi$^{®}$69	No subsidies	22,162.8	9152.8	13,010.0	101.1
	Arrosi$^{®}$G 1a	No subsidies	23,564.4	9486.8	14,077.6	109.3
	Arrosi$^{®}$240	No subsidies	24,966.0	9152.8	15,813.2	122.8

4. Discussion

4.1. Economic Evaluation

The results shown in Table 5 indicate that all materials had similar yields to PE film, but the trend is that some of the biodegradable materials obtain higher yields, confirming previous evidences such as that of [32] who reported higher pepper yields with similar biodegradable materials compared to PE.

Total costs and net margins (Table 6) in the PE situations are quite similar, with an increase of 0.11% in the costs when considering landfill and 0.18% when plastic is recycled. These results suggest that the cost of waste treatment and recycling do not significantly affect final profitability. This contrasts strongly with the widespread perception among farmers that waste management is costly in terms of time and money. Our estimations support the authorities' efforts to hold farmers responsible for the wastes they generate in their activities until the end of their cycle.

However, given that there are no significant yield differences between materials, it is important to note that subsidies would be insufficient to compensate for the extra cost of the material if identical yields were obtained, with the only exception of Ecovio® and Sphere®. This result is maintained even taking into account the total recycling cost. Therefore, the current level of subsidies (35%) does not seem to be a strong enough incentive for all the biodegradable materials to be adopted by farmers. An alternative to the current system should provide for compensation to cover the difference in cost with regard to PE. Calculations show that the rate of subsidy should be 50.1% for Mater-Bi® and 37.6% for Bioflex® to assure these options to be as profitable as PE. When the total cost of recycling is considered, then the necessary subsidy would reach 48.7% for Mater-Bi® and 35.9% for Bioflex®.

With regard to paper mulches, although their costs are between 5.5% and 9.3% higher than PE, they obtain higher net margins due to the influence of savings on field conditioning operations and higher yields. Arrosi®240 is the best option among paper mulches, with increases in net margin by 22.8%. Once again, this result is highly dependent on the higher yields. When yields are considered the same as obtained by PE, then the over-cost of paper materials is not compensated by savings in waste management costs. In this case, the percentage of subsidies needed to make them as profitable as the PE option would be 48.2% for Mimgreen®, 45.1% for Arrosi® 69 and Arrosi® 240, and 58.6% for Arrosi® G1a.

In summary, although six of the eight materials evaluated as alternatives to PE have proved to be more profitable, only two of them (Ecovio® and Sphere®) are good potential alternatives from an economic point of view under the current subsidies received despite their higher market price. Two main reasons explain this result: first, because they achieve crop yields similar to PE, and secondly, because they save waste treatment costs that compensate their higher market prices. Biodegradable plastics benefit from public support to compensate for part of the rise in market prices but the results show that the current subsidies system does not guarantee the profitability of all the materials analyzed. In fact, the most expensive materials (Mater-Bi® and Bioflex®) are not good economic alternatives when the yields are the same as PE. Similarly, [1] showed that the use of biodegradable mulches with tomato crop in different localities was only profitable in certain specific locations and with some materials.

Interestingly, two of the evaluated biodegradable films (Ecovio® and Sphere®) are good economic alternatives to PE under the current public payment system. This contrasts with the widespread use of the PE, which probably comes from its low cost in comparison with biodegradable materials. By contrast, our calculations show that biodegradable films can be better alternatives in the short-term even in the case of no waste management. The net margins when using these biodegradable materials are even better when recycling is considered mandatory. Of course, there may be other non-economic reasons that may inhibit broader adoption of bio materials and papers. Breakdown during the growing season and fragments of mulches after tillage may be aesthetically displeasing to farmers and consumers thus inhibiting their adoption. In the case of papers, it may also exist a negative perception linked to the greater discomfort for their installation beyond the cost of time that has been included in our calculations.

4.2. Environmental Implications of the Use of Plastic Films and Papers

In addition to the short-term economic considerations, other environmental aspects related to the use of mulching materials should be taken into account. It is necessary to emphasize the increasing problems caused in the environment by the plastics. For example, [33] indicated that the presence of PE in horticultural soils in Argentina can represent around 10% of the soil and [34] affirmed that the amount of plastic waste in an average vegetable field of China could reach 317.4 kg ha^{-1}. Although no similar data have been found for Europe, there is strong evidence that the presence of plastic residues also affect the soil quality. For example, [13] reported that amounts of residual mulch films of 320 kg ha^{-1} could interfere in tomato crop yields, causing decreases by 5.9% in yields. It has been demonstrated that this effect on the soil's productive capacity increases with the concentration of plastic particles in the soil. This evidence is a further argument in favor of making the complete management of waste mandatory for farmers, and therefore a strong support for the use of other biodegradable materials.

However, it should be remembered that there is a growing number of studies warning of the consequences of the use of many of the so-called "biodegradable" materials, as they do not degrade completely in soil. A recent study of [23] hypothesized the case where a farmer tills all the biodegradable mulch at the end of the crop cycle into the soil. The standard method tests applied to plastics (ASTM D5988 and ISO 17556) consider a degradation rate of 90% biodegradation rate within to 2 years; considering this, 45% of this plastic will remain in the field during the first year. After the second year, a 10% of the first year plastic will probably remain in soil and the plastic from the second application with its 10% remaining to the third year. If this 10% is assumed never to degrade, then it will accumulate every year. The authors hypothesize that 350 kg ha^{-1} of non-degradable plastic will represent 6.45% decreased yield on the fifth year of using biodegradable films and tilling them at the end of the crop season. Unfortunately, there is no standard method to measure the rate of degradation after incorporation in the soil and the percentages could be very variable.

In the case of some of our tested materials, some evidences are reported in literature. [35] established that Bioflex® material lost 73% of their initial weight after 145 days after soil incorporation (DASI), while Sphere lost only 42% in the same period. On the other hand, Mater-Bi® generated fragments of a wide range of sizes (up to 2664 mm^2) which maybe will interfere with tillage, another aspect to take into consideration. By contrast, the paper Mimgreen® presented the smallest fragments and surface after 200 DASI.

With regard to paper mulches, no waste management has to be implemented and no accumulation of waste in the soil is expected to interfere with the crop, so, in principle, their effects are likely to be less harmful than plastics. However, papers are insufficiently explored until now and their environmental effects in the long-term and these advantages have to be proven. If these advantages are verified, then the papers should be eligible for public support.

5. Conclusions

The extensive use of PE mulching materials owes to their lower market prices compared to biodegradable materials. However, our results show that the inclusion of the costs of waste management and recycling is crucial for a proper evaluation of the economic profitability of different options in the short-term. The inclusion of such costs under the current Spanish legislation only increases the costs by 9.5 € ha^{-1} with respect to the no waste management scenario and 15.5 € ha^{-1} if total recycling cost is considered. These increases represent a reduction in the final net margin of 0.1%. This is supporting the mandatory measures for farmers to assume the costs of waste management and recycling.

Economic consideration of current Spanish government support of biodegradable mulching materials allows us to affirm that only two materials (Ecovio® and Sphere®) are profitable alternatives to PE when the same yield is considered. Despite the saving in costs of field conditioning with regard to PE, the high market prices of biodegradable and paper materials are not compensated with the current level of subsidies, thus impeding their adoption in fields. An increase in subsidies rates of up to 50.1% would allow all biodegradable films to be better alternatives than PE.

Although no fully conclusive evidence has been found on the environmental effects of long-term use of the specific materials analyzed, the consideration of soil quality effects supports measures towards mandatory full recycling of waste and for the use of biodegradable and paper materials. Correct assessment of environmental damages of materials would require other types of field experiments than those conducted here. These data could be included in a long-term economic model based on the analysis of the net present value of discounted future social costs (economic plus environmental damages) and benefits (yield gains and reduced environmental damages). In addition, an adequate evaluation must take into account that subsidies provide an economic incentive for the adoption of bio-materials, but also an opportunity cost to society, thus a proper design must be ensured.

Finally, although this study refers to field trials with pepper crops, the results may be representative of the open-air growing conditions for other summer horticultural crops under similar climatic conditions, mainly in the Ebro Valley, where mulches are often used.

Author Contributions: A.I.M., G.P. and A.C. conceived and designed the field experiments; A.I.M., G.P. and A.C. performed the field experiments; A.I.M. compiled and analyzed the field data; Y.M. analyzed the economic results; All the authors wrote the paper.

Funding: This work has been financed by the INIA Projects RTA2011-00104-C04-00 and RTA2017-00082-00-00 funded by the Spanish Ministry of Agriculture and Fisheries, Food and Environment, and ECO2016-75927-R funded by the Spanish Ministry of Economics, Industry and Competitiveness.

Acknowledgments: We thank JA Alins, F Arrieta, and D Redondo for their help and valuable contribution. We thank Verso Paper Corporation, Fábrica de papeles crepados Arrosi S.A. Fábrica de papeles crepados, BASF S.E., FKuR Kunststoff GmbH, Mimcord S.A., Novamont S.p.A., Oerlemans Plastics BV, Stora-EnsoFinland and Sphere Group Spain S.L. for generously providing the materials and the market prices.

References

1. Cirujeda, A.; Aibar, J.; Anzalone, A.; Martín-Closas, L.; Meco, R.; Moreno, M.M.; Pardo, A.; Pelacho, A.M.; Rojo, F.; Royo-Esnal, A.; et al. Biodegradable mulch instead of polyethylene for weed control of processing tomato production. *Agron. Sustain. Dev.* **2012**, *32*, 889–897. [CrossRef]

2. Cirujeda, A.; Anzalone, A.; Aibar, J.; Moreno, M.M.; Zaragoza, C. Purple nutsedge (*Cyperus rotundus*) control with paper mulch in processing tomato. *Crop Prot.* **2012**, *39*, 66–71. [CrossRef]

3. Zhang, G.S.; Zhang, X.X.; Hu, X.B. Runoff and soil erosion as affected by plastic mulch patterns in vegetable field at Dianchi Lake's catchment, China. *Agric. Water Manag.* **2013**, *122*, 20–27. [CrossRef]

4. Abdul Kader, M.; Senge, M.; Abdul Mojid, M.; Nakamura, K. Mulching type-induced soil moisture and temperature regimes and water use efficiency of soybean under rain-fed condition in central Japan. *Int. Soil Water Conserv. Res.* **2017**, *5*, 302–308. [CrossRef]

5. Miles, C.; Wallace, R.; Wszelaki, A.; Martin, J.; Cowan, J.; Walters, T.; Inglis, D. Deterioration of potentially biodegradable alternatives to black plastic mulch in three tomato production regions. *HortScience* **2012**, *47*, 1270–1277.

6. Summers, C.G.; Stapleton, J.J. Management of corn leafhopper (Homoptera: *Cicadellidae*) and corn stunt disease in sweet corn using reflective mulch. *J. Econ. Entomol.* **2002**, *95*, 325–330. [CrossRef] [PubMed]

7. Ingman, M.; Santelmann, M.V.; Tilt, B. Agricultural water conservation in China: Plastic mulch and traditional irrigation. *Ecosyst. Health Sustain.* **2015**, *1*, art.12. [CrossRef]

8. Jabran, K.; Ullah, E.; Hussain, M.; Farooq, M.; Zaman, U.; Yaseen, M.; Chauhan, B.S. Mulching improves water productivity, yield and quality of fine rice under water-saving rice production systems. *J. Agron. Crop Sci.* **2015**, *201*, 389–400. [CrossRef]

9. Ghimire, S.; Miles, C. Dimensions and Costs of Polyethylene, Paper and Biodegradable Plastic Mulch. Washington State University. Extension Factsheet 2016. Available online: http://plasticulture.wsu.edu/there-are-no-further-articles-in-this-category/publications/ (accessed on 5 April 2018).

10. Shogren, R.L.; Hochmuth, R.C. Field evaluation of watermelon grown on paper-polymerized vegetable oil mulches. *HortScience* **2004**, *39*, 1588–1591.

11. Immirzi, B.; Santagata, G.; Vox, G.; Schettini, E. Preparation, characterization and field-testing of a biodegradable sodium alginate-based spray mulch. *Biosyst. Eng.* **2009**, *102*, 416–472. [CrossRef]

12. Steinmetz, Z.; Wollmann, C.; Schaefer, M.; Buchmann, C.; David, J.; Tröger, J.; Muñoz, K.; Frör, O.; Schaumann, G.E. Plastic mulching in agriculture. Trading short-term agronomic benefits for long-term soil degradation? *Sci. Total Environ.* **2016**, *550*, 690–705. [CrossRef] [PubMed]

13. Zou, X.; Niu, W.; Liu, J.; Li, Y.; Liang, B.; Gou, L.; Guan, Y. Effects of residual mulch film on the growth and fruit quality of tomato (*Lycopersicon esculentum* Mill.). *Water Air Soil Pollut.* **2017**, *228*, 71. [CrossRef]

14. Díaz-Pérez, J.C. Bell pepper (*Capsicum annum* L.) grown on plastic film mulches: Effects on crop microenvironment, physiological attributes and fruit yield. *HortScience* **2010**, *45*, 1196–1204.

15. PlasticsEurope. Plastics—The Facts 2017. An Analysis of European Plastics Production, Demand and Waste Data. Available online: https://www.plasticseurope.org/application/files/5715/1717/4180/Plastics_the_facts_2017_FINAL_for_website_one_page.pdf (accessed on 2 February 2018).

16. Lamont, W.J. Plastics: Modifying the microclimate for the production of vegetable crops. *HortTechnology* **2005**, *15*, 477–481. [CrossRef]

17. MAPAMA, Ministerio de Agricultura, Pesca, Alimentación y Medio Ambiente. Plan Estatal Marco de Gestión de Residuos (PEMAR) 2016–2022. Madrid. Available online: http://www.magrama.gob.es/es/calidad-y-evaluacion-ambiental/planes-y-estrategias/pemaraprobado6noviembrecondae_tcm7-401704.pdf (accessed on 15 November 2016).

18. Liu, E.K.; He, H.Q.; Yan, C.R. 'White revolution' to 'white pollution': Agricultural plastic film mulch in China. *Environ. Res. Lett.* **2014**, *9*, 091001. [CrossRef]

19. Scarascia-Mugnozza, G.; Sica, C.; Russo, G. Plastic materials in European agriculture: Actual use and perspectives. *J. Agric. Eng.* **2011**, *42*, 15–28. [CrossRef]

20. Gross, R.A.; Kalra, B. Biodegradable polymers for the environment. *Science* **2002**, *297*, 803–807. [CrossRef]

21. Brault, D.; Stewart, K.A. Growth, development and yield of head lettuce cultivated on paper and polyethylene mulch. *HortScience* **2002**, *37*, 92–94.

22. Martín-Closas, L.; Bach, M.A.; Pelacho, A.M. Biodegradable mulching in an organic tomato production system. In *Proceedings of the International Symposium on Sustainability*; International Society for Horticultural Science: Seoul, Korea, 2008; pp. 267–273.

23. Miles, C.; DeVetter, L.; Ghimire, S.; Hayes, D.G. Suitability of biodegradable plastic mulches for organic and sustainable agricultural production systems. *HortScience* **2017**, *52*, 10–15. [CrossRef]

24. Moreno, M.M.; Cirujeda, A.; Aibar, J.; Moreno, C. Soil thermal and productive response of biodegradable mulch materials in a processing tomato (*Lycopersicon esculentum* Mill.) crop. *Soil Res.* **2016**, *54*, 207–215. [CrossRef]

25. FAOSTAT. Available online: http://www.fao.org/faostat/en/#home (accessed on 12 March 2018).

26. Vázquez, N.; Huete, J.; Pardo, A.; Suso, M.L.; Tobar, V. Use of soil moisture sensors for automatic high frequency drip irrigation in processing tomato. XXVIII International Horticultural Congress on Science and Horticulture for People. *Acta Hortic.* **2011**, *922*, 229–235. [CrossRef]

27. MAPAMA, Ministerio de Agricultura, Pesca, Alimentación y Medio Ambiente. Encuesta Sobre Superficies y Rendimientos de Cultivos (ESYRCE) (2017). Available online: http://www.mapama.gob.es/es/estadistica/temas/estadisticas-agrarias/agricultura/esyrce/ (accessed on 30 March 2018).

28. Marí, A.I. Acolchados y su Efecto Sobre el control y la Biología de la Juncia (*Cyperus rotundus* L.). Ph.D. Thesis, University of Lérida, Lérida, Spain. waiting for the public defense.

29. Gartraud. *Serres Maraîchers: La Gestion de Déchets Solides. Jornada Técnica: Buenas Prácticas Agrarias en Horticultura*; DARP, Plan Anual de Transferencia Tecnológica: Viladecans, Barcelona, Spain, 2004.

30. MAPAMA, Ministerio de Agricultura, Pesca, Alimentación y Medio Ambiente. Estudios de Costes de Explotaciones Agrarias (ECREA) (2018). Available online: https://www.mapama.gob.es/gl/ministerio/servicios/analisis-y-prospectiva/ECREA_Informes-Agricolas.aspx (accessed on 25 May 2018).

31. Gobierno de Aragón. Departamento de Desarrollo Rural y Sostenibilidad (2018) O.C.M. de Frutas y Verduras. Available online: http://www.aragon.es/portal/site/GobiernoAragon/menuitem.a5eff4c9604087b9bad5933754a051ca/?vgnextoid=c88c8e1b65b9b210VgnVCM100000450a15acRCRD&idTramite=1006 (accessed on 20 February 2018).

32. Lahoz, I.; Macua, J.I.; Cirujeda, A.; Aibar, J.; Marí, A.I.; Pardo, G.; Suso, M.; Pardo, A.; Moreno, M.M.; Moreno, C.; et al. Influencia del acolchado biodegradable en la producción de pimiento. In *Proceedings of the XIII Jornadas del Grupo de Horticultura, I Jornadas del Grupo de Alimentación y Salud*; Sociedad Española de Ciencias Hortícolas: Logroño, Spain, 2014; pp. 203–208.

33. Ramos, L.; Berenstein, G.; Hughes, E.A.; Zalts, A.; Montserrat, J.M. Polyethylene film incorporation into the horticultural soil of small periurban production units in Argentina. *Sci. Total Environ.* **2015**, *523*, 74–81. [CrossRef]

34. Zhao, F.J.; Ma, Y.; Zhu, Y.G.; Tang, Z.; McGrath, S.P. Soil contamination in China: Current status and mitigation strategies. *Environ. Sci. Technol.* **2014**, *49*, 750–759. [CrossRef]
35. Moreno, M.M.; González-Mora, S.; Villena, J.; Campos, J.A.; Moreno, C. Deterioration pattern of six biodegradable, potentially low-environmental impact mulches in field conditions. *J. Environ. Manag.* **2017**, *200*, 490–501. [CrossRef]

Article

Using Flaming as an Alternative Method to Vine Suckering

Luisa Martelloni *, Michele Raffaelli, Christian Frasconi, Marco Fontanelli, Andrea Peruzzi and Claudio D'Onofrio

Department of Agriculture, Food and Environment, University of Pisa, Via del Borghetto 80, 56124, Pisa, Italy; michele.raffaelli@unipi.it (M.R.); christian.frasconi@unipi.it (C.F.); marco.fontanelli@unipi.it (M.F.); andrea.peruzzi@unipi.it (A.P.); claudio.donofrio@unipi.it (C.D.)
* Correspondence: lmartelloni@agr.unipi.it; Tel.: +39-050-2218-966

Received: 19 February 2019; Accepted: 19 March 2019; Published: 21 March 2019

Abstract: Suckering is the process of removing the suckers that grapevine trunks put out in the spring. Suckering by hand is costly and time consuming and requires constant bending down, getting up and making repetitive motions. The mechanical removal of suckers with rotating scourges can damage the vine plants. Chemical suckering is a limiting factor for wine grape growers interested in sustainable and/or organic agriculture. The aim of this research was to test flaming as an alternative method to vine suckering. A three-year experiment was conducted on a 10-year-old Sangiovese vine (775 Paulsen rootstock). The treatments consisted of flame suckering at different phenological stages, hand-suckering and a no-suckered control. Data on the number of suckers, grape yield components, and grape composition were collected and analysed. The results showed that flaming significantly reduced the initial number of suckers. This effect on the suckers was highest when the main productive shoots of the vines were at the 18-19 BBCH growth stage. Flame-suckering did not affect grape yield components and grape composition. Future studies could investigate the simultaneous use of flaming for both suckering and weed control.

Keywords: grapevine; no-chemical; organic agriculture; sucker removal; *Vitis vinifera* (L.); thermal

---

1. Introduction

Suckers are nonbearing shoots that grow in the spring from latent buds on grapevine (*Vitis vinifera* L.) trunks [1]. Sucker growth can lead to excess vegetation, increase the possibility of attack from pathogens and alter the fruit/shoot ratio [2]. Moreover, suckers can cause problems during vineyard management operations, such as soil tillage, weed removal, mechanical harvest, and pest and disease control [3]. To overcome these problems suckers are removed during grapevine cultivation and this process is known as suckering. The right time for suckering is when they are not yet lignified. Waiting longer causes the suckers to become lignified, harden, which are then more difficult to remove. Suckering in spring also prevents the development of resprouting basal buds [4].

Traditionally, suckering was done by hand, however this is costly and time consuming because it requires constant bending down, getting up and making repetitive motions [5]. Hand suckering requires an operating time ranging from a minimum of 20 h ha^{-1} to a maximum of 60-70 h ha^{-1}, depending on the operating conditions [6]. The mechanical removal of suckers by scourges is widely employed, however this is generally stressful on young plants, which can be damaged by rotating scourges [2]. Chemical suckering with traditional herbicides or synthetic growth regulators is also widely used [2,7], however the use of synthetic chemicals is forbidden for organic wine grape growers.

Flaming could be a viable nonchemical alternative to remove the not yet lignified spring suckers. The high temperature of the flame denaturises the plant proteins of green tissues, without burning, and thus desiccates them [8]. Flame-suckering could be useful for organic viticulture, which has received increased interest by grapevine growers in the recent decades.

Flaming is currently used to control weeds in heat-tolerant herbaceous and horticultural crops [9–12], however, to the best of our knowledge, there has been no research using flaming to remove suckers from grapevines. This research tests the effects of flaming to remove the suckers. Grape yield components and grape composition were also recorded.

2. Materials and Methods

2.1. Experimental Set Up

A three-year experiment (2016, 2017 and 2018) was conducted on a 10-year-old Sangiovese vine (clone BF-30) grafted on 775 Paulsen rootstock. The farm (Tenuta Ceppaiano, Castellani Spa) was located in Tuscany, Italy (43°35'51.6'' N 10°32'13.8'' E). The vineyard training system was spurred cordon. The cordons were 80 cm height. The distance between each vine on the row was 80 cm, and between the rows was 2.10 m, for a density of 5952 plants per ha. The soil was loam (40% sand, 34% silt, 26% clay, 1% organic matter, pH = 7). Figure 1 reports the monthly-cumulated rainfall and monthly average temperatures recorded during the three-year experiment. Fertilization consisted of the application of an organic-mineral fertilizer in January 2016 and 2017 (10N-5P-14K and 8N-16P-24K, respectively), and calcium nitrate (15.5N–0P–0K) in January 2018. Sixteen, eight and eleven chemical treatments, against *Plasmopara viticola* (Berk. & M.A. Curtis) and *Uncinula necator* (Schwein.) Burrill, were applied from April to August in 2016, 2017 and 2018, respectively. One chemical treatment against *Lobesia botrana* (Schiff. et Den.) was applied in June in 2016 and 2018. The vineyard was not irrigated.

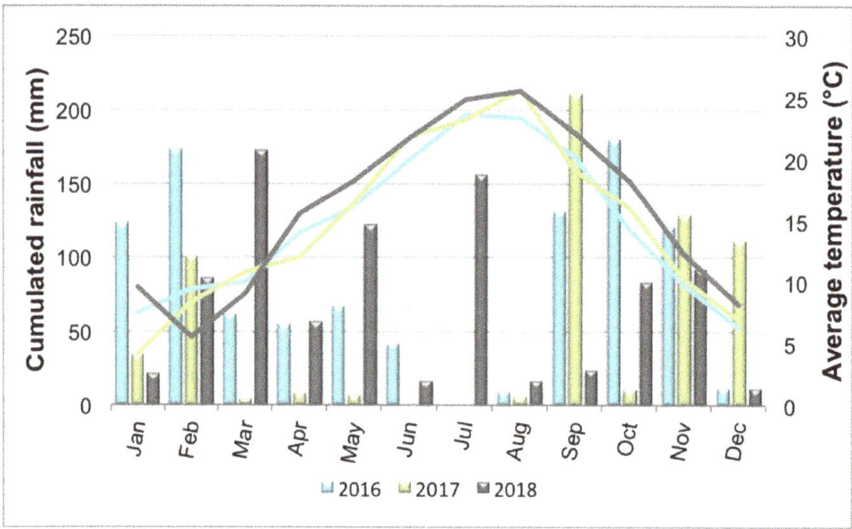

Figure 1. Monthly cumulated rainfall and monthly average temperatures (January 2016–December 2018) recorded by the meteorological station in Siberia, Crespina-Lorenzana (Pisa, Italy) (43°35'31.2'' N 10°32'38.4'' E) [13].

The flaming machine used for suckering was the PFV-600 model (Officine Mingozzi, Ferrara, Italy) [14] (Figure 2). A mobile horizontal frame supports the burners, which are placed in two rows in a staggered position. The inclination of the burner rows can be adjusted based on the height of the suckers. The burners were cylindrical with air-intake and operated in gaseous phase (Figure 2) [15]. The flaming machine was coupled with a SAME Frutteto 100 (Same, Treviglio, Bergamo, Italy) tractor.

Figure 2. The flaming machine PFV-600 (Officine Mingozzi, Ferrara, Italy) coupled with a SAME Frutteto 100 (SAME, Treviglio, Italy) tractor.

The flaming machine was supplied with liquefied petroleum gas (LPG). The LPG consumption at the pressure of 0.2 MPa was 17.64 kg h^{-1}. The working width was 1.05 m (half of the 2.10 m inter-row space) and the forward speed was 3 km h^{-1}. The machine distributed 55.90 kg ha^{-1} of LPG. The LPG dose actually applied to the suckers (within the intra-row space of 0.30 m) was 195.65 kg ha^{-1}. This LPG dose was chosen because it was deemed effective to devitalize suckers on the basis of previous experiments where flaming was used to devitalize weeds and cover crops [10,11,16–19].

Each year, the first flaming was applied in the spring when the vine plants showed the most developed suckers at the 12–13 BBCH growth stage (two–three unfolded leaves) and the main productive shoots at the 15–16 BBCH growth stage (five–six unfolded leaves) [20] (21 April in 2016 and 2017, 2 May in 2018) (Figures 3 and 4) (Supplementary Materials). The second flaming was applied when the nonflamed plants showed the most developed suckers at the 15–16 BBCH growth stage and the main shoots at the 18-19 BBCH growth stage (eight–nine unfolded leaves) (5 May in 2016 and 2017, 16 May in 2018). Hand suckering was conducted on the same date as the first flaming. The control was not suckered in the spring, but lignified suckers were manually removed during the winter pruning. The suckers that remained on the plants after the flaming and hand suckering, or that had resprouted, were also removed during the winter pruning.

Figure 3. Flame suckering applied on 21 April 2017 at the 13 BBCH sucker growth stage.

Figure 4. Flame suckering applied on 21 April 2016 at the 12–13 BBCH sucker growth stage.

The experimental design was a randomized complete block design. Five adjacent vineyard rows were selected, and each row was divided into five 16-m-long blocks (one for each treatment). Treatments were: (1) flaming applied once only when the most developed suckers were at the 12–13 BBCH growth stage (treatment "FlamingA"), (2) flaming applied once only when the most developed suckers were at the 15–16 BBCH growth stage (treatment "FlamingB"), (3) flaming applied twice, the first time when the most developed suckers were at the 12–13 BBCH growth stage, and the second time at the same date as FlamingB (treatment "FlamingC"), (4) hand suckering when the most developed suckers were at the 12–13 BBCH growth stage (treatment "Hand"), and (5) nonsuckered plants (treatment "Control") (Table 1).

Table 1. Date of suckering treatments, and the growth stage of the most developed suckers at the time of suckering.

Treatment	Year	Date of Application	Suckers Growth Stage
FlamingA	2016	21 April	12–13 BBCH
FlamingB	2016	5 May	15–16 BBCH
FlamingC	2016	21 April and 5 May	12–13 BBCH on 21 April, 15–16 BBCH on 5 May
Hand	2016	21 April	12–13 BBCH
FlamingA	2017	21 April	12–13 BBCH
FlamingB	2017	5 May	15–16 BBCH
FlamingC	2017	21 April and 5 May	12–13 BBCH on 21 April, 15–16 BBCH on 5 May
Hand	2017	21 April	12–13 BBCH
FlamingA	2018	2 May	12–13 BBCH
FlamingB	2018	16 May	15–16 BBCH
FlamingC	2018	2 May and 16 May	12–13 BBCH on 2 April, 15–16 BBCH on 16 May
Hand	2018	2 May	12–13 BBCH

2.2. Data Collection

Each year, data were always collected in relation to the same five vine plants at the centre of each block for a total of 25 replicates for each treatment. The persistence of suckers after treatments was evaluated by counting the number of suckers at four different times: (1) immediately before the first flaming, (2) two weeks after FlamingA, (3) three weeks after FlamingA, and one week after FlamingB, and (4) seven weeks after FlamingA, and five weeks after FlamingB.

In September, at the harvest, all the clusters of each replicate were counted and weighed together in order to evaluate the yield. The average cluster weight (g cluster^{-1}) was calculated by dividing the yield by the number of clusters. The average berry weight (g berry^{-1}) was calculated by averaging the weight of 50 berries randomly picked from the clusters of each replicate.

Immediately after harvest, the berries were placed in hermetically sealed plastic bags and stored in a cooler at 4 °C to preserve their characteristics. The berries were then crushed and the juice filtered through cheesecloth to determine total soluble solids, pH and tartaric acid following standard methods (European Commission Regulation (EC) No. 2676/90). Total soluble solids (Brix) were determined at 20 °C using an ATC digital refractometer (Tekcoplus, Hong Kong, China); pH was measured using a Hanna H18519N electronic pH-meter (Hanna Instruments, Padova, Italy); and tartaric acid was determined by acid-base titration using sodium hydroxide (0.1 N) to an endpoint pH of 8, with values expressed as tartaric acid (g L^{-1}).

Flaming machine performance parameters and costs were calculated. The field efficiency (i.e., the ratio of the theoretical field time and the total time spent in the field) was computed by referring to a hypothetical area of 10,000 m^2 (30.00 m wide and 333.33 m long). The theoretical field time is the time the machine is effectively operating at an optimum forward speed and performing over its full width of action. The total time for conducting the operation was calculated by summing the machine adjustment time (including plugging and unplugging), the theoretical field time, the turning time, and the time to refuel the tractor and/or replace empty LPG tanks. However, the travelling time back and forth the field was not included. The total cost per use was calculated by summing the fixed and variable costs for the flaming machine coupled with a SAME Frutteto 100, following a standard methodology for cost determination [21]. The rate of depreciation was determined considering a purchase price of €46,445 for the SAME Frutteto 100, and € 12,139 for the flaming machine. The economic lifetime considered was 12 years for the tractor, and 10 years for the flaming machine. The repairing and maintenance factor was 80% for the tractor, and 75% for the flaming machine. The labour costs for the tractor driver was 15 € h^{-1}, and the LPG cost was 2.25 € kg^{-1}.

2.3. Statistical Analysis

Data normality was assessed using the Shapiro–Wilk test. Other tests consisted of the Student t-test to verify that the mean error was not significantly different to zero, the Breusch–Pagan test for homoscedasticity and the Durbin–Watson test for autocorrelation.

Data on the number of suckers were modelled in a generalized linear mixed model using the extension package lmerTest (Tests in Linear Mixed Effects Models) [22] of R software [23]. The log transformation was assessed. The treatments and data collection dates were the fixed factors. The random factors (replicates and data collection dates) were assessed as longitudinal data (repeated measures) to obtain a correlated random effect for intercept and slope. Data were analysed separately each year. The analysis of deviance was run. The package emmeans (Estimated Marginal Means, aka Least-Squares Means) [24] was used to compute the least squares means, standard errors, inverse transformed values, and confidence intervals.

Yield components and grape composition data were modelled in a linear mixed model using the extension package lmerTest [22] of R software [23]. Treatments and years were the fixed factor, and replicates and years were the random factors. The analysis of variance was run. The package lmerTest [22] was used to compute the least squares means and standard errors.

Pairwise comparisons between estimated least squares means were computed by estimating the 95% confidence interval of the difference between the least squares means (Equation (1)):

$$\text{CI (difference)} = (x_1 - x_2) \pm 1.96 \sqrt{(\text{SE}_{x_1})^2 + (\text{SE}_{x_2})^2} \tag{1}$$

where (x_1) is the mean of the first value, (x_2) is the mean of the second value, $(\text{SE}x_1)$ is the standard error of (x_1), and $(\text{SE}x_2)$ is the standard error of (x_2) [25].

If the resulting 95% confidence interval (CI) of the difference between values did not cross the zero value, the null hypothesis that the compared values were not different was rejected.

3. Results

3.1. Number of Suckers

The analysis of deviance showed in all the years that the number of suckers was influenced by the type of treatment, the data collection date, and their interaction ($p < 0.001$, respectively). For each year of the experiment, the least squares means and the standard errors of the number of suckers log transformed are reported in Table 2. Inverse transformed values and lower and upper confidence intervals are reported in Table 3. The number of suckers before the first flaming was similar between treatments in all years. In 2016, all the flaming treatments and hand suckering significantly reduced the number of suckers compared to the initial number of suckers. In 2017 and 2018 only FlamingA (suckers flamed only on 21 April 2017 and 2 May 2018) did not significantly reduce the initial number of suckers. In the control plot, the number of suckers was similar to that before the start of the experiment in all the years (Table 2).

Table 2. Least squares means of the number of suckers per plant log transformed and standard errors (SE) as affected by different types of suckering methods and data collection date in the three-year experiment. These data are useful for computing all pairwise comparisons.

Date	Year	Log (No. of Suckers) (±SE)				
		FlamingA	FlamingB	FlamingC	Hand	Control
21 April	2016	1.38 (0.11)	1.40 (0.11)	1.43 (0.13)	1.58 (0.16)	1.37 (0.12)
5 May	2016	0.65 (0.17)	1.40 (0.12)	0.77 (0.18)	0.44 (0.28)	1.41 (0.13)
12 May	2016	0.67 (0.17)	0.41 (0.18)	−0.97 (0.41)	0.57 (0.26)	1.42 (0.13)
9 June	2016	0.68 (0.16)	0.59 (0.17)	−0.56 (0.34)	0.57 (0.26)	1.45 (0.12)
21 April	2017	1.04 (0.13)	0.98 (0.14)	1.11 (0.13)	1.38 (0.19)	0.98 (0.15)
5 May	2017	0.80 (0.15)	1.10 (0.14)	0.73 (0.15)	−0.60 (0.46)	1.00 (0.15)
12 May	2017	0.98 (0.14)	0.50 (0.17)	−0.01 (0.21)	−0.42 (0.42)	1.04 (0.14)
9 June	2017	0.98 (0.14)	0.58 (0.17)	0.11 (0.20)	−0.01 (0.35)	1.10 (0.14)
2 May	2018	0.88 (0.19)	1.07 (0.17)	0.88 (0.19)	1.01 (0.18)	0.95 (0.18)
16 May	2018	0.28 (0.25)	1.12 (0.17)	0.08 (0.28)	−0.41 (0.36)	1.01 (0.18)
23 May	2018	0.40 (0.24)	0.22 (0.26)	−0.41 (0.36)	−0.19 (0.32)	1.10 (0.18)
20 June	2018	0.51 (0.23)	0.40 (0.24)	−0.09 (0.31)	−0.01 (0.29)	1.18 (0.17)

FlamingA = plants flamed only on 21 April in 2016 and 2017, 2 May in 2018. The most developed suckers were at the 12–13 BBCH growth stage. FlamingB = plants flamed only on 5 May in 2016 and 2017; 16 May in 2018. The most developed suckers were at the 15–16 BBCH growth stage. FlamingC = plants flamed on 21 April and 5 May in 2016 and 2017; 2 May and 16 May 2018. Hand = plants hand suckered on 21 April 2016 and 2017; 2 May in 2018. Control = no suckered plants.

Table 3. Inverse transformed values (from the log scale) and lower and upper confidence intervals (LCI, UCI) of the least squares means of the number of suckers per plant as affected by different type of suckering methods and data collection date in the three-year experiment.

Date	Year	No. of Suckers (LCI, UCI)				
		FlamingA	FlamingB	FlamingC	Hand	Control
21 April	2016	3.97 (3.17, 4.98)	4.04 (3.25, 5.03)	4.17 (3.26, 5.34)	4.88 (3.59, 6.63)	3.93 (3.08, 5.01)
5 May	2016	1.92 (1.39, 2.66)	4.07 (3.25, 5.11)	2.16 (1.52, 3.05)	1.55 (0.90, 2.65)	4.10 (3.20, 5.25)
12 May	2016	1.96 (1.42, 2.71)	1.51 (1.06, 2.16)	0.38 (0.17, 0.85)	1.76 (1.06, 2.94)	4.28 (3.37, 5.45)
9 June	2016	1.97 (1.43, 2.72)	1.80 (1.30, 2.50)	0.57 (0.30, 1.10)	1.77 (1.07, 2.93)	4.28 (3.37, 5.46)
21 April	2017	2.84 (2.19, 3.67)	2.66 (2.03, 3.48)	3.04 (2.37, 3.90)	3.97 (2.74, 5.74)	2.66 (2.01, 3.52)
5 May	2017	2.24 (1.67, 2.99)	3.00 (2.30, 3.91)	2.08 (1.54, 2.81)	0.55 (0.22, 1.35)	2.72 (2.04, 3.62)
12 May	2017	2.66 (2.01, 3.50)	1.65 (1.18, 2.32)	0.99 (0.65, 1.50)	0.66 (0.29, 1.51)	2.84 (2.13, 3.78)
9 June	2017	2.68 (2.04, 3.51)	1.79 (1.29, 2.48)	1.12 (0.76, 1.66)	0.99 (0.50, 1.96)	3.00 (2.27, 3.95)
2 May	2018	2.41 (1.66, 3.50)	2.91 (2.07, 4.09)	2.41 (1.66, 3.50)	2.74 (1.93, 3.89)	2.59 (1.81, 3.72)
16 May	2018	1.33 (0.81, 2.18)	3.07 (2.20, 4.29)	1.08 (0.62, 1.87)	0.66 (0.33, 1.33)	2.76 (1.94, 3.92)
23 May	2018	1.49 (0.93, 2.39)	1.24 (0.74, 2.08)	0.66 (0.33, 1.33)	0.83 (0.44, 1.55)	3.01 (2.13, 4.24)
20 June	2018	1.66 (1.06, 2.60)	1.49 (0.93, 2.39)	0.91 (0.50, 1.66)	0.99 (0.56, 1.77)	3.26 (2.35, 4.52)

FlamingA = plants flamed only on 21 April in 2016 and 2017, 2 May in 2018. The most developed suckers were at the 12–13 BBCH growth stage. FlamingB = plants flamed only on 5 May in 2016 and 2017; 16 May in 2018. The most developed suckers were at the 15–16 BBCH growth stage. FlamingC = plants flamed on 21 April and 5 May in 2016 and 2017; 2 May and 16 May 2018. Hand = plants hand suckered on 21 April 2016 and 2017; 2 May in 2018. Control = no suckered plants.

In 2016, on 5 May, the number of suckers in FlamingA and FlamingC was significantly lower compared to FlamingB (not yet flamed) and the control, and similar to the hand-suckered plants. On 12 May, all treatments had a lower number of suckers compared to the control. The number of suckers was also lower in FlamingC compared to other treatments, whereas, FlamingA, FlamingB and the hand-suckered plants showed a statistically similar number of suckers. On 9 June, the differences in the number of suckers between treatments were the same as the previous data collection date (four weeks before) (Table 2).

In 2017, on 5 May, the number of suckers in the flamed plants was similar to the nonflamed plants. The hand-suckered plants showed a significantly lower number of suckers compared to the other treatments. On 12 May, the number of suckers in FlamingC, FlamingB and hand-suckered treatments was similar and significantly lower compared to FlamingA and the control (FlamingA and the control were statistically similar). On 9 June, the number of suckers in FlamingA and FlamingB treatments was similar. FlamingA was also similar to the control, whereas FlamingC, FlamingB and the hand-suckering were similar and showed a lower number of suckers compared to the control (Table 2).

In 2018, on 16 May, the number of suckers on the flamed plants was similar to the hand-suckered plants and significantly lower compared to FlamingB and the control. On 23 May, the number of suckers was higher in the control and similar between all other treatments. Flaming and hand-suckering treatments were similar and showed a lower number of suckers compared to the control also on 20 June (Table 2).

In all years, for each flaming treatment and hand suckering, no significant increase in the number of suckers was observed comparing the data collection date after the treatment (two weeks for Flaming A, FlamingC and hand suckering, and one week for FlamingB) and the last data collection date. This thus suggests that in this time period no resprouting of suckers occurred. In all the years, the resprouting of suckers after treatment occurred only during the two weeks between the start of the experiment and the first data collection (i.e., from 21 April to 5 May in 2016 and 2017, and from 2 May and 16 May in 2018) (Table 2).

3.2. Yield Components and Grape Composition

For each model, all the p-values of the analysis of variance are reported in Table 4. The suckering method (treatment) and the treatment:year interaction were not significant, whereas the year was significant for all the dependent variables analysed.

Table 4. Analysis of variance for yield, average cluster weight, average berry weight, total soluble solids, pH and tartaric acid, and analysis of deviance (type II Wald chi-square test) for the number of clusters.

Variable	p-Values		
	Treatment	Year	Treatment:Year
Yield (kg plant^{-1})	0.928	<0.001	0.743
No. of clusters (cluster plant^{-1})	0.979	<0.001	1.000
Average cluster weight (g cluster^{-1})	0.435	<0.001	0.324
Average berry weight (g berry^{-1})	0.464	<0.001	0.874
Total soluble solids (°Brix)	0.674	<0.001	0.998
pH	0.861	<0.001	0.685
Tartaric acid (g L^{-1})	0.430	<0.001	0.730

Table 5 reports the least squares means and standard errors of the yield, clusters number (log transformed and inverse transformed values with the lower and upper 95% confidence interval), average cluster weight and average berry weight as affected by type of treatment and the year. Within the same year, the yield, number of clusters, and the average weights of the clusters and berries were similar between treatments, suggesting that flame-suckering did not affect yield components.

In the flame-suckered plots, the yield decreased significantly in 2017 compared to 2016, and in 2018 compared to 2017. In the hand-suckered and control plots, the yield was similar in 2016 compared to 2017, whereas in 2018 was lower compared to 2016 and 2017. The average berry weight in 2016 was similar to 2017, whereas it was higher in 2018 compared to 2016 and 2017 for all the treatments. Except for the hand-suckered plants, where the number of clusters per plant was similar in 2016 and 2017, for the other treatments, the number of clusters increased from 2016 to 2017. The number of clusters decreased significantly from 2017 to 2018 for all the treatments. The average cluster weight was significantly lower in 2017 and 2018 compared to 2016 and was similar in 2017 and 2018 for all the treatments.

Table 5. Least squares means and standard errors of the yield, clusters number (log transformed) average cluster weight and average berry weight as affected by the type of treatment and the year. Inverse transformed values and lower and upper 95% confidence intervals (LCI, UCI) of clusters number log transformed are reported.

Variable	Year	Treatment				
		FlamingA	FlamingB	FlamingC	Hand	Control
Yield (kg plant^{-1}) ($\pm$SE)	2016	3.46 (0.24)	3.57 (0.24)	3.36 (0.24)	3.42 (0.39)	3.12 (0.24)
Yield (kg plant^{-1}) ($\pm$SE)	2017	2.52 (0.24)	2.46 (0.24)	2.73 (0.25)	2.93 (0.39)	2.88 (0.24)
Yield (kg plant^{-1}) ($\pm$SE)	2018	1.80 (0.35)	1.50 (0.335)	1.56 (0.335)	1.91 (0.35)	1.51 (0.335)
log[Clusters (no. plant^{-1})] ($\pm$SE)	2016	2.24 (0.07)	2.23 (0.07)	2.22 (0.07)	2.21 (0.11)	2.20 (0.07)
log[Clusters (no. plant^{-1})] ($\pm$SE)	2017	2.42 (0.06)	2.42 (0.06)	2.41 (0.06)	2.38 (0.10)	2.40 (0.06)
log[Clusters (no. plant^{-1})] ($\pm$SE)	2018	1.98 (0.11)	1.94 (0.11)	1.93 (0.11)	1.94 (0.115)	1.94 (0.11)
Clusters (no. plant^{-1}) (LCI, UCI)	2016	9.42 (8.25, 10.77)	9.34 (8.17, 10.68)	9.21 (8.05, 10.54)	9.08 (7.28, 11.32)	9.05 (7.90, 10.36)
Clusters (no. plant^{-1}) (LCI, UCI)	2017	11.30 (10.00, 12.77)	11.21 (9.92, 12.68)	11.14 (9.82, 12.63)	10.85 (8.85, 13.29)	11.01 (9.72, 12.46)
Clusters (no. plant^{-1}) (LCI, UCI)	2018	7.25 (5.81, 9.04)	6.97 (5.61, 8.64)	6.88 (5.54, 8.55)	6.94 (5.53, 8.70)	6.97 (5.61, 8.64)
Average cluster weight (g cluster^{-1}) ($\pm$SE)	2016	357.09 (17.37)	389.24 (17.37)	368.13 (17.37)	379.66 (28.33)	344.07 (17.37)
Average cluster weight (g cluster^{-1}) ($\pm$SE)	2017	235.17 (17.37)	216.80 (17.37)	226.27 (17.72)	278.07 (28.33)	258.70 (17.37)
Average cluster weight (g cluster^{-1}) ($\pm$SE)	2018	239.02 (25.15)	215.32 (24.13)	217.47 (24.13)	273.51 (25.39)	221.21 (24.13)
Average berry weight (g berry^{-1}) ($\pm$SE)	2016	1.55 (0.09)	1.58 (0.09)	1.71 (0.09)	1.68 (0.09)	1.60 (0.09)
Average berry weight (g berry^{-1}) ($\pm$SE)	2017	1.68 (0.15)	1.92 (0.15)	1.66 (0.15)	1.82 (0.15)	1.63 (0.15)
Average berry weight (g berry^{-1}) ($\pm$SE)	2018	2.18 (0.13)	2.36 (0.13)	2.35 (0.13)	2.26 (0.13)	2.22 (0.13)

FlamingA = plants flamed only on 21 April in 2016 and 2017, 2 May in 2018. The most developed suckers were at the 12–13 BBCH growth stage. FlamingB = plants flamed only on 5 May in 2016 and 2017; 16 May in 2018. The most developed suckers were at the 15–16 BBCH growth stage. FlamingC = plants flamed on 21 April and 5 May in 2016 and 2017; 2 May and 16 May 2018. Hand = plants hand suckered on 21 April 2016 and 2017; 2 May in 2018. Control = no suckered plants.

Least-squares means and standard errors of the total soluble solids, pH and tartaric acid as affected by the type of treatment and the year are reported in Table 6. The values of total soluble solids, pH and tartaric acid were statistically similar between treatments in all the years. In 2016, the total soluble solids and pH values were significantly lower compared to values estimated in 2017 and 2018 for all the treatments. The amount of tartaric acid was significantly higher in 2016 compared to 2017 and 2018 for all the treatments. The total soluble solids, pH and tartaric acid content were similar in 2017 and 2018 for all the treatments. As total soluble solids increase in the berries, the juice pH rises and the tartaric acid declines.

Table 6. Least-squares means and standard errors of the total soluble solids, pH, and tartaric acid as affected by the type of treatment and the year.

Variable	Year	Treatment				
		FlamingA	FlamingB	FlamingC	Hand	Control
Total soluble solids (°Brix) (±SE)	2016	20.26 (0.59)	20.38 (0.59)	20.74 (0.59)	20.74 (0.59)	20.70 (0.54)
Total soluble solids (°Brix) (±SE)	2017	22.75 (0.86)	22.90 (0.86)	23.70 (0.86)	23.60 (0.86)	22.70 (0.86)
Total soluble solids (°Brix) (±SE)	2018	21.80 (0.65)	22.13 (0.65)	22.25 (0.65)	22.38 (0.65)	22.43 (0.65)
pH (±SE)	2016	3.13 (0.03)	3.11 (0.03)	3.12 (0.03)	3.18 (0.03)	3.12 (0.03)
pH (±SE)	2017	3.41 (0.06)	3.48 (0.06)	3.43 (0.06)	3.47 (0.06)	3.47 (0.06)
pH (±SE)	2018	3.47 (0.05)	3.46 (0.05)	3.54 (0.05)	3.44 (0.05)	3.54 (0.05)
Tartaric acid (g L^{-1})	2016	6.59 (0.20)	6.50 (0.20)	6.36 (0.20)	6.15 (0.20)	6.40 (0.18)
Tartaric acid (g L^{-1})	2017	5.48 (0.31)	4.65 (0.31)	5.18 (0.31)	5.25 (0.31)	5.20 (0.31)
Tartaric acid (g L^{-1})	2018	4.91 (0.22)	4.63 (0.22)	4.80 (0.22)	4.95 (0.22)	4.74 (0.22)

FlamingA = plants flamed only on 21 April in 2016 and 2017, 2 May in 2018. The most developed suckers were at the 12–13 BBCH growth stage. FlamingB = plants flamed only on 5 May in 2016 and 2017; 16 May in 2018. The most developed suckers were at the 15–16 BBCH growth stage. FlamingC = plants flamed on 21 April and 5 May in 2016 and 2017; 2 May and 16 May 2018. Hand = plants hand suckered on 21 April 2016 and 2017; 2 May in 2018. Control = no suckered plants.

3.3. Flaming Machine Performance and Costs

Flaming machine performance and costs are reported in Table 7. It should be pointed out that a large amount of the variable costs for conducting flaming was due to the high cost of the LPG in Italy (2.25 € kg^{-1}), and that flaming may be less expensive in countries where the LPG (or propane) would costs less (e.g., in the USA the propane cost is equivalent to 0.48 € kg^{-1}).

Table 7. Flaming machine performance and costs estimation. The machine was used coupled with a SAME Frutteto 100 tractor.

Performance	
Forward speed (km h^{-1})	3.00
Working width (m)	1.05
Theoretical field capacity (ha h^{-1})	0.31
Theoretical field time (h) *	3.17
Turning time (h) *	0.23
Time to refuel the tractor and/or replace empty LPG tanks (h) *	0.19
Machine adjustment time (includes plugging and unplugging) (h)	0.25
Total time (h) *	3.84
Field efficiency *	0.83
Effective field capacity (ha h^{-1}) *	0.26
Costs	
Tractor cost per hour (€ h^{-1}) *	28.15
Tractor cost per use (€ ha^{-1}) *	108.10
Flaming machine cost per hour (€ h^{-1}) *	43.23
Flaming machine cost per use (€ ha^{-1}) *	137.23
Total cost per hour (€ h^{-1}) *	71.38
Total cost per use (€ ha) *	245.33

* Time to conduct the operation in a hypothetical area of 10,000 m^2 (30.00 m wide and 333.33 m long).

4. Discussion

Flame suckering reduced the number of suckers observed before the start of the experiment each year (Table 2). The date of flaming influenced the final number of suckers. In fact, in 2017 and 2018, the suckers removed with flaming on 21 April 2017 and 2 May 2018 led to a significant resprouting of suckers compared to the flaming two weeks later on 5 May 2017 and 16 May 2018 (or applied on both dates), which resulted in a nonsignificant reduction in the initial number of suckers (Table 2). Flame suckering was more effective if not carried out too early, because although the suckers were more developed (15–16 BBCH growth stage *vs* 12–13 BBCH growth stage), and therefore would seem more difficult to devitalize, sucker resprouting is more difficult due to the more developed main grape productive shoots (18-19 BBCH growth stage *vs* 15–16 BBCH growth stage).

In fact, a vine is a 'acrotonic branching' plant due to the distal position of shoots, irrespectively of its size relative to the parent stem [26]. More developed shoots exploit their higher sink strength, and absorb more nutrients [27,28]. Apical dominance of *Vitis vinifera L.* increases with increasing temperature [29]. This explains why, over time, there was a significant decrease in the resprouting of basal buds. The amount of resprouting, in the first two weeks after the start of the experiment, was the same as the resprouting in the hand-suckered plots, where suckers were completely removed during the operation. However, two weeks later the number of suckers was no longer zero, but on average there were two suckers per plant in 2016 and one sucker per plant in 2017 and 2018. This amount, added to approximately one sucker per plant which flaming did not devitalize, led to the nonsignificant decrease in the number of suckers compared to the initial number when FlamingA was applied in 2017 and 2018. There was no resprouting in the control plants because new buds only sprout when buds (e.g., the suckers) have been removed from the plant [4].

In the three-year experiment, the yield components and grape composition were similar among treatments, suggesting that flaming did not damage the grapevine plants, or modify the quality of the grapes. Byrne and Howell [3] found that sucker removal increased the yield per vine. In our experiment, the low plant potential fertility due to the low number of buds per vine, the tendency to an excess of total leaf surface per plant in relation to the quantity of grape production, and the few suckers per plant (a maximum of around four suckers per control plant in 2016), although statistically higher compared to the flame- and hand-suckering, was probably not enough to negatively affect the production of grapes.

Although the number of suckers per plant did not influence the yield, and suckers were manually removed during the winter pruning (as per general vineyard procedure), suckering during the spring is nevertheless required. This is because the presence of suckers during the vine growing season can cause problems during vineyard management such as weed, insect, and disease control, and mechanical harvest [3]. Moreover, postponing the suckering of a high number of suckers until winter pruning would significantly increase the cost of manually removing the excessively developed and lignified shoots, and most of all, would increase the development of basal buds capable of resprouting [4]. Finally, the majority of vine disease pathogens infect via winter wounds, because winter is associated with high rainfall, which increases inoculum availability, increasing the risk of sucker wounds acting as portals for grapevine trunk pathogen infections [30,31].

The reduction in yield from 2016 to 2018 cannot be attributed to the suckering, because the reduction was found in all treatments indiscriminately. It was probably the result of the dry year in 2017 (Figure 1), which led to a lower potential fertility for 2018, resulting in a lower number of clusters. The higher average berry weight estimated in 2018 was probably because of the lower number of clusters and the higher rainfall occurring during the 2018 growing season compared to 2016 and 2017 (Figure 1). Concerning grape composition, as the total soluble solids increase in the berries, the juice pH rises and the tartaric acid declines.

5. Conclusions

Almost always, flame-suckering led to a significant reduction in suckers in each year of application, even if suckers flamed one time at the 12–13 BBCH growth stage (FlamingA) did not show a reduction in the initial number of suckers in 2017 and 2018 (Table 2). On the other hand, flaming when suckers were at the 15–16 BBCH growth stage (when vine main shoots were at the 18-19 BBCH) (FlamingB and FlamingC) were the most effective flaming treatments, thus suggesting that delaying flame suckering reduces the number of suckers that resprout. Double-flaming (FlamingC) significantly reduced the number of suckers compared to hand-suckering alone in 2016, suggesting that the time when suckers are removed is probably more important than the number of flaming treatments. Moreover, conducting flame-suckering only once reduces by half the total cost per use. Flame suckering thus seems to be a valid alternative to the use of chemicals for organic growers. In addition, the number of suckers removed by flaming was similar to that of manual suckering. This is an important outcome, because hand suckering is very expensive (e.g., about 10 € h^{-1} for a labour time up to 60–70 h ha^{-1}). Future studies could investigate the simultaneous use of flaming for both suckering and weed control, in order to provide an economic, sustainable alternative to a chemical approach for organic grapevine growers.

Supplementary Materials: The following are available online at http://www.mdpi.com/2073-4395/9/3/147/s1, Video S1: Flame suckering applied on 21 April 2016 at the 12–13 BBCH sucker growth stage.

Author Contributions: Conceptualization, M.R., L.M., C.F., M.F., A.P. and C.D.; methodology, M.R., L.M., C.F., M.F., A.P. and C.D.; validation, L.M., M.R., C.F., M.F., A.P., and C.D.; formal analysis, L.M., M.R., C.F., M.F., A.P. and C.D.; investigation, L.M., M.R., C.F., M.F., A.P. and C.D.; resources, L.M., M.R., C.F., M.F., A.P. and C.D.; data curation, L.M.; writing—original draft preparation, L.M.; writing—review and editing, L.M., M.R., C.F., M.F., A.P. and C.D.; visualization, L.M., M.R., C.F., M.F., A.P. and C.D.; supervision, L.M., M.R., C.F., M.F., A.P. and C.D.; project administration, M.R., L.M., C.F., and C.D.

Funding: This research received no external funding.

Acknowledgments: This study was self-financed by the Department of Agriculture, Food and Environment of the University of Pisa (Pisa, Italy). The authors would like to thank Adrian Wallwork (e4ac.com) for providing the English language editing of this paper; Valentina Panicucci from the Department of Agriculture, Food and Environment of University of Pisa (Pisa, Italy) for her contribution to the 2016 experiment; Piergiorgio Castellani from Castellani Spa (Pontedera, Pisa, Italy) for hosting the trials; and Alessandro Moretto from Castellani Spa, Paolo Belluomini and Piero Puntoni from the Department of Agriculture, Food and Environment of University of Pisa for their technical support.

Conflicts of Interest: The authors declare no conflicts of interest.

References

1. Hellman, E.W. Grapevine structure and function. In *Oregon Viticulture*; Hellman, E.W., Ed.; Oregon State University Press: Corvallis, OR, USA, 2003; pp. 5–19.
2. Dolci, M.; Galeotti, F.; Curir, P.; Schellino, L.; Gay, G. New 2-naphthyloxyacetates for trunk sucker growth control on grapevine (*Vitis vinifera* L.). *Plant Growth Regul.* **2004**, *44*, 47–52. [CrossRef]
3. Byrne, M.E.; Howell, G.S. Initial response of Baco noir grapevines to pruning severity, sucker removal, and weed control. *Am. J. Enol. Vitic.* **1978**, *29*, 192–198.
4. Fregoni, M. Chapter X. La potatura della vite. In *Viticoltura di Qualità*; Edizioni l'Informatore Agrario: Verona, Italy, 1999; pp. 377–492. (In Italian)
5. Kang, F.; Wang, H.; Pierce, F.J.; Zhang, Q.; Wang, S. Sucker detection of grapevines for targeted spray using optical Sensors. *Trans. ASABE* **2012**, *55*, 2007–2014. [CrossRef]
6. Paliotti, A.; Poni, S.; Silvestroni, O. Interventi in verde. Spollonatura. In *La nuova Viticoltura. Innovazioni Tecniche per Modelli Produttivi Efficienti e Sostenibili*; Edagricole: Milano, Italy, 2015; p. 141.
7. Ahmedullah, M.; Wolfe, W.H. Control of sucker growth on *Vitis vinifera* L. cultivar Sauvignon Blanc with naphthalene acetic acid. *Am. J. Enol. Vitic.* **1982**, *33*, 198–200.
8. Mojžiš, M. Energetic requirements of flame weed control. *Res. Agric. Eng.* **2002**, *48*, 94–97.
9. Knezevic, S. Flame weeding in corn, soybean and sunflower. In Proceedings of the 8th International Conference on Information and Communication Technologies in Agriculture, Food and Environment (HAICTA 2017), Chania, Crete Island, Greece, 21–24 September 2017; pp. 390–394.

10. Martelloni, L.; Fontanelli, M.; Frasconi, C.; Raffaelli, M.; Peruzzi, A. Cross-flaming application for intra-row weed control in maize. *Appl. Eng. Agric.* **2016**, *32*, 569–578. [CrossRef]

11. Martelloni, L.; Fontanelli, M.; Frasconi, C.; Raffaelli, M.; Pirchio, M.; Peruzzi, A. A combined flamer-cultivator for weed control during the harvesting season of asparagus green spears. *Span. J. Agric. Res.* **2017**, *15*, e0203. [CrossRef]

12. Neilson, B.D.; Bruening, C.A.; Stepanovic, S.; Datta, A.; Knezevic, S.; Gogos, G. Design and field testing of a combined flaming and cultivation implement for effective weed control. *Appl. Eng. Agric.* **2017**, *33*, 43–54.

13. Regione Toscana. Settore Idrologico Regionale. Ricerca dati. Centro Funzionale Regionale di Monitoraggio Meteo—Idrologico. Available online: http://www.sir.toscana.it/ricerca-dati (accessed on 12 February 2019).

14. Officine Mingozzi. Technical Info. Available online: http://www.pirodiserbo.it/dati_tecnici_pfv.pdf (accessed on 12 February 2019).

15. Raffaelli, M.; Frasconi, C.; Fontanelli, M.; Martelloni, L.; Peruzzi, A. LPG burners for weed control. *Appl. Eng. Agric.* **2015**, *31*, 717–731. [CrossRef]

16. Frasconi, C.; Martelloni, L.; Fontanelli, M.; Raffaelli, M.; Marzialetti, P.; Peruzzi, A. Thermal weed control in Photinia × fraseri "Red Robin" container nurseries. *Appl. Eng. Agric.* **2017**, *33*, 345–356. [CrossRef]

17. Frasconi, C.; Martelloni, L.; Antichi, D.; Raffaelli, M.; Fontanelli, M.; Peruzzi, A.; Benincasa, P.; Tosti, G. Combining roller crimpers and flaming for the termination of cover crops in herbicide-free no-till cropping systems. *PLoS ONE* **2019**, *14*, e0211573. [CrossRef] [PubMed]

18. Martelloni, L.; Caturegli, L.; Frasconi, C.; Gaetani, M.; Grossi, N.; Magni, S.; Peruzzi, A.; Pirchio, M.; Raffaelli, M.; Volterrani, M.; et al. Use of flaming to control weeds in 'Patriot' hybrid bermudagrass. *Horttechnology* **2018**, *28*, 843–850. [CrossRef]

19. Peruzzi, A.; Martelloni, L.; Frasconi, C.; Fontanelli, M.; Pirchio, M.; Raffaelli, M. Machines for nonchemical intra-row weed control in narrow and wide-row crops: A review. *J. Agric. Eng.* **2017**, *48*, 57–70. [CrossRef]

20. Lorenz, D.H.; Eichhorn, K.W.; Bleiholder, H.; Klose, R.; Meier, U.; Weber, E. Growth stages of the grapevine: Phenological growth stages of the grapevine (*Vitis vinifera* L. ssp. vinifera)—Codes and descriptions according to the extended BBCH scale. *Aust. J. Grape Wine Res.* **1995**, *1*, 100–103. [CrossRef]

21. Hunt, D. Chapter 4. Costs. In *Farm Power and Machinery Management*; Waveland Press: Long Grave, IL, USA, 2001; pp. 75–77.

22. Kuznetsova, A.; Brockhoff, P.B.; Christensen, R.H.B. lmerTest: Tests in Linear Mixed Effects Models. R Package Version 2.0-32. Available online: https://CRAN.R-project.org/package=lmerTest (accessed on 12 February 2019).

23. R Core Team. *R: A Language and Environment for Statistical Computing*; R Foundation for Statistical Computing: Vienna, Austria. Available online: https://www.R-project.org/ (accessed on 12 February 2019).

24. Lenth, R. Emmeans: Estimated Marginal Means, aka Least-Squares Means. R Package Version 1.2. 2018. Available online: https://CRAN.R-project.org/package=emmeans (accessed on 12 February 2019).

25. Knezevic, A. Overlapping Confidence Intervals and Statistical Significance. 2008. Available online: https://www.cscu.cornell.edu/news/statnews/stnews73.pdf (accessed on 12 February 2019).

26. Bell, A.D. *Plant Form: An Illustrated Guide to Flowering Plant Morphology*; Oxford University Press: New York, NY, USA, 1991.

27. Williams, L. Growth of 'Thompson Seedless' grapevines. I. Leaf area development and dry weight distribution. *J. Am. Soc. Hortic. Sci.* **1987**, *112*, 325–330.

28. Miller, D.P.; Howell, G.S.; Flore, J.A. Effect of shoot number on potted grapevines. Canopy development and morphology. *Am. J. Enol. Vitic.* **1996**, *47*, 244–250.

29. Buttrose, M.S. Some effects of light intensity and temperature on dry weight and shoot growth of grape-vine. *Ann Bot.* **1968**, *32*, 753–765. [CrossRef]

30. Lecomte, P.; Bailey, D.J. Studies on the infestation by Eutypa lata of grapevine spring wounds. *Vitis* **2011**, *50*, 35–51.
31. Makatini, G.H. The Role of Sucker Wounds as Portals for Grapevine Trunk Pathogen Infections. Master's Thesis, Plant Pathology, MScAgric, Stellenbosch University, Stellenbosch, South Africa. Available online: http://hdl.handle.net/10019.1/86599 (accessed on 13 February 2019).

Article

Cover Crop Effectiveness Varies in Cover Crop-Based Rotational Tillage Organic Soybean Systems Depending on Species and Environment

Laura Vincent-Caboud [1,*], Léa Vereecke [2], Erin Silva [2] and Joséphine Peigné [1]

[1] Department of Agroecology and Environment, ISARA-Lyon (member of the University of Lyon), 23 rue Jean Baldassini, F-69364 Lyon CEDEX 07, France; jpeigne@isara.fr
[2] Department of Agronomy, University of Wisconsin-Madison, Madison, WI 53706, USA; vereecke@wisc.edu (L.V.); emsilva@wisc.edu (E.S.)
* Correspondence: lavincent.caboud@gmail.com; Tel.: +33-042-785-8573

Received: 4 May 2019; Accepted: 14 June 2019; Published: 18 June 2019

Abstract: Organic farming relies heavily on tillage for weed management, however, intensive soil disturbance can have detrimental impacts on soil quality. Cover crop-based rotational tillage (CCBRT), a practice that reduces the need for tillage and cultivation through the creation of cover crop mulches, has emerged as an alternative weed management practice in organic cropping systems. In this study, CCBRT systems using cereal rye and triticale grain species are evaluated with organic soybean directly seeded into a rolled cover crop. Cover crop biomass, weed biomass, and soybean yields were evaluated to assess the effects of cereal rye and winter triticale cover crops on weed suppression and yields. From 2016 to 2018, trials were conducted at six locations in Wisconsin, USA, and Southern France. While cover crop biomass did not differ among the cereal grain species tested, the use of cereal rye as the cover crop resulted in higher soybean yields (2.7 t ha^{-1} vs. 2.2 t ha^{-1}) and greater weed suppression, both at soybean emergence (231 vs. 577 kg ha^{-1} of weed biomass) and just prior to soybean harvest (1178 vs. 1545 kg ha^{-1}). On four out of six sites, cover crop biomass was lower than the reported optimal (<8000 kg ha^{-1}) needed to suppress weeds throughout soybean season. Environmental conditions, in tandem with agronomic decisions (e.g., seeding dates, cultivar, planters, etc.), influenced the ability of the cover crop to suppress weeds regardless of the species used. In a changing climate, future research should focus on establishing flexible decision support tools based on multi-tactic cover crop management to ensure more consistent results with respect to cover crop growth, weed suppression, and crop yields.

Keywords: weed management; organic farming; mulch; weed dynamic; cereal grain cover crop; roller-crimper

---

1. Introduction

Worldwide, land under certified organic production reached 698 million hectares in 2017 [1]. Across the global organic land base, the production of organic soybean [*Glycine max* (L.) Merr.] is increasing, with 429,621 ha under production in 2017 [1]. With more than 39,996 ha of organic soybean grown in 2014, the United States is the third largest producer of organic soybean [1–3]. In recent years, the European market is also rapidly expanding, with 72,710 ha of organic soybean production in 2016 [1,4]. Within Europe, France leads organic soybean production with 24,615 ha.

Improved weed management and increased crop productivity have emerged as two main levers to facilitate the expansion of organic soybean acreage and meet the production demand [5,6]. As the prohibition of most synthetic substances is included in global organic regulatory frameworks, alternative techniques have been developed to manage weeds, including mechanical cultivation, strategic crop rotation, and the use of cover crops [7–10]. For most organic farmers, soil tillage is necessary to manage weeds, prepare the seedbed, and incorporate organic inputs [11]. However, intensive soil disturbance may decrease soil quality (e.g., reducing organic matter, increasing soil erosion, etc.), thereby raising concerns on the sustainability of organic farming practices [12].

To maintain soil fertility, organic farmers are encouraged by the Food and Agriculture Organization of the United Nations (FAO) to reduce soil tillage, improve soil coverage and diversify crop rotation [13–15]. Among all the techniques developed to reduce tillage, organic cover crop-based rotational tillage systems (organic CCBRT) has emerged as a practice of great interest. These systems reduce tillage through the establishment of cash crops into high residue cover crops terminated with a roller-crimper [16–19]. The cover crop mulch remains on the soil surface until cash crop harvest, preventing weed emergence, and thus eliminating the need for mechanical weed management, maintaining soil quality while reducing labor and fuel consumption. In addition to creating a physical barrier which reduces weed emergence, an additional mechanism of weed suppression includes the competition of the cover crop with weeds for water, nutrients, and light [20,21]. Further, weed control may also be enhanced through allelopathic compounds released by the cover crop, which can inhibit weed germination [20,22–26].

Currently, reduced tillage practices implemented within conventional row crop systems are highly dependent on the use of chemical herbicides [17,27–29]. Growing concerns about the detrimental impacts of herbicides and the increasing occurrence of herbicide-resistant weeds have stimulated research interest for CCBRT in both organic and conventional production systems, especially in the United States where this technique has seen significant growth over the past decade [19,30,31]. The technique is less developed in Europe, but farmers' interest in preserving soil quality is increasing, as shown by a European survey conducted in 2012 on organic conservation practices [32].

Previous research has shown that effective weed control can be achieved through CCBRT until crop harvest if the cover crop biomass reaches from 8000 to 10,000 kg ha^{-1} according to conditions (e.g., climate, weed infestation, weed species) before termination [16,33]. Cover crop species selection also serves as a fundamental tool to (1) optimize cover crop biomass, (2) inhibit weed germination through the release of allelopathic compounds and (3) ensure adequate termination of the cover crop with a roller-crimper [34–36]

Some cereal grain cover crops perform well in CCBRT systems with soybean cash crops, including cereal rye (*Secale cereale* L.), triticale (*x Triticosecale Wittmack*), barley (*Hordeum vulgare* L.), oat (*Avena sativa* L.)], and winter wheat (*Trticum vulgare* L.) [37,38]. Their main advantages over legume species are the high biomass production and consistent termination with a roller-crimper.

Among the cereal grain cover crops, cereal rye has consistently superior performance in the organic CCBRT system, producing high amounts of dry matter and reaching anthesis (Zadoks stage 69) [36,37], the stage of maturity necessary for mechanical termination, earlier than other cereals [19,21,39,40]. Cereal rye has also exhibited a high degree of allelopathy, inhibiting weed seed germination [41,42]. Incomplete mechanical termination of cereal rye in organic CCBRT may result in volunteer cereal rye plants in subsequent phases of the crop rotation, which results in contamination of following crops with rye grain, affecting both quality and yields of subsequent crops [18,40,43,44]. Thus, in recent years, triticale and barley, species with lesser propensity to produce volunteer plants, have been explored as alternative cover crops to rye. Additionally, the more prostrate growth habit and wider leaves characteristic of these species may provide greater light interception, improving early season weed control [39,45]. However, a dearth of references exists on the comparative performances of different cereal species in organic CCBRT systems.

While previous studies have demonstrated the ability of cereal rye cover crops to suppress weeds, the success of the CCBRT technique remains highly variable across years and location [18,46]. Investigation of the performance of organic CCBRT systems over a broad range of pedoclimatic conditions with the comparison of some cereal grain cover crops is needed to understand the reasons for failures and achieve more consistent success. Alternative cereal grain species such as triticale could provide similar results than cereal rye and provide benefits for facing soil and climate condition to reach consistent cover crop performance. The objective of this study was to examine the performance of two cover crop species used in combination with soybean in an organic CCBRT system under different pedoclimatic conditions through a multi-site experiment over two years: (1) In the Upper-Midwestern USA and (2) Southeastern France. This study aimed (i) to determine which cover crop species leads to the highest soybean success rate and (ii) determine the drivers of variability in cover crop performance, weed control and soybean yields observed in different pedoclimatic conditions.

2. Materials and Methods

2.1. Site Description

The trials were conducted on certified organic land at two locations in 2017 and four in 2018, located in the upper Midwestern U.S. and in Southern France. The US sites are characterized by short growing season with high seasonal rainfall, cold winter conditions, and warm summer temperatures, as compared to the European sites which were defined by a more temperate climate, with consistent cool conditions and lower precipitation.

Site A is the University of Wisconsin Arlington Agricultural Research Station (UW-AARS) in Arlington, WI, USA. The four other locations are in Southern France Rhône-Alpes region, with site B in Drôme, site C in Northwestern Isère, site D in Ain and site E in Northeastern Isère. Soil types and climates are presented in Table 1. At Arlington (site A), fields have been certified organic since 2009 and were under alfalfa cover crop from 2014 through 2016. The organic CCBRT system trial was initiated in 2017 and relies on the common four-year rotation practiced in the upper Midwestern U.S., including, corn, soybean, fallow, and small grain [19]. In Southern France (sites B, C, D, and E), annual trials were implemented in the typical crop rotation practiced by farmers under organic grain system which is based on similar crops rotation as encountered in the upper Midwestern U.S. (i.e., winter wheat, corn, soybean, alfalfa). At sites B and E, reduced tillage was practiced throughout the prior 10 year period, while the historical management practices at C and D sites relied on traditional tillage. Sites B, C, and D have been certified organic for 13–27 years, while sites E has been managed organically for three years.

Table 1. Description of the six experimental sites (soil and climate conditions).

Site	Year	Previous Crop	Soil Type	Organic Matter (%)	pH	Climate (Location)	Irrigation System (Yes/No)
17-Arl. A1 18-Arl. A2	2017 2018	Alfalfa Corn	Plano silt loam	3.7	7.3	Humid continental climate, 889 mm, 9.45 °C (UWAARS, 43°18′N, 89°21″ E, 315 masl)	No
17-Frce B	2017	Winter wheat	Loamy sand	2.6	7.8	Mediterranean climate, 835 mm, 12.1 °C (45°00′40.2″N 4°59′07.1″E)	Yes
18-Frce C	2018	Winter wheat	Fine loam clay	4.9	8.4	Oceanic and temperate climate, 877 mm, 11.3 °C (45°40′51.3″N 5°32′13.9″E)	No
18-Frce D	2018	Alfalfa	Loamy sand	2.7	8.5	Semi-continental climate with Mediterranean influence, 785 mm, 11.5 °C (45°49′10.9″N 5°02′05.6″E)	yes
18-Frce E	2018	Winter wheat	Fine loam	1.6	7.5	Warm temperate climate, 797 mm, 11.5 °C (45°35′09.9″N 4°55′29.3″E)	No

2.2. Experimental Design and Crop Management

At each location, two cover crop species were compared (cereal rye and triticale) using a randomized complete block design with four replications. The detailed field operations are presented in Table 2. Site A (Arlington, WI, USA) was a 0.48 ha field with 67 m × 9 m sub-plots. The French sites (B, C, D and E) were 0.23 ha fields with 24 × 12 m sub-plots. Winter rye, ('*Aroostook*' (site A), '*Dukato*' (site C, D, E), '*Ovid*' (site B)) and winter triticale ('*NF426GT*' (site A), '*Vuka*' (site B, C, D, E)) were planted at the end of summer or early fall of 2016 and 2017 (Table 2). Different 3 m wide drills were used depending on the location (site A-Model 750, John Deere, Moline, IL, site B and E-Sulky Master, site C-Saphir 7/400-DS 125, Lemken). On site D, the drill was a 4 m wide Vitasem 402 A, Pottinger. Planting depth was standardized at 2.5 cm.

Roller-crimpers of different widths, weight and manufacturers were used to terminate the cover crops (site A 4.6 m, 1360 kg, I and J Manufacturing, Gap, PA, sites B, C and E 3 m, 1400 kg, University of Lyon 1, Rhône-Alpes region, France, site D-6 m, 3300 kg, FACA, Sky Agriculture). Soybeans were planted and cover crops were terminated when the latter reached 50% to 100% anthesis (Zadoks growth stage 65–69) both years, thus resulting in different soybean planting dates depending on year.

Soybeans were planted with a 4.6 m wide conservation tillage planter in Wisconsin (site A) (Model 1750 Max Emerge Plus, Conservation Tillage, John Deere, Moline, IL), a 6 m wide no-till drill on site C and D (Easydrill W 6000, Sky Agriculture), a 3-m wide no-till drill on site E (Easydrill 3000 Fertisem, Sky Agriculture), and a 4 m wide planter on site B (Maxima 2 TI M, Kuhn) (see Table 2 for row spacing). Crimping and planting were performed the same day in two separate passes across the field, except for 18-Frce E site where both crimping and planting were performed as a one-pass operation.

2.3. Data Collection

Weather data for site A was obtained from a meteorological station located at UW-AARS (from 2016 to 14 November 2017) and the Michigan State University Enviroweather Service (from 15 November 2017 to 2018). In France, individual stations were used for each site: Valence-Chabeuil (site B), Bourgoin (site C), Lyon-Bron (site D) and Reventin (site E). Weather data was collected from the fall of 2016 to the fall of 2018.

Cover crop biomass (in kg of dry matter.ha^{-1}) was determined by collecting aboveground biomass in four randomized quadrats per plots before cover crop termination (quadrat size 0.5 m × 0.75 m in France, 0.5 m × 0.5 m in Wisconsin). The samples were dried at 80 °C until constant weight. Cover crop height was also recorded on 20 randomized plants per plots.

Weed biomass (in kg of dry matter.ha^{-1}) was determined by collecting aboveground biomass in three randomized 0.5 m × 0.75 m quadrats per plots centered on the row at two different dates: (i) Date 1 in the summer (July or August) and (ii) Date 2 in the fall (September) prior to soybean harvest. Samples were dried at 80 °C until constant weight. Weed species were identified on each plot of each site in the summer (Date 1) to document the dominant weed species.

Soybean stands were determined by counting emerged plants on three randomized four linear meters portions of the rows within three weeks after planting. Soybean aboveground biomass was estimated at the flowering stage (between R3 and R5 soybean stage) on three randomized two linear meters per plot. Soybean height was measured on 15 randomized soybean plant per plots at the mid-flowering stage (between R3 and R5 soybean stage). In France, to estimate soybean yields, soybean aboveground biomass and soybean grain weight were measured on two linear meters. Samples were replicated three times per plot or 12 times per cover crop species (cereal rye and triticale). At Arlington (site A), yields were measured using a 4.6 m wide combine (Gleaner, AGCO) in 2017 and a two-row plot combine in 2018.

Table 2. Field operations at the different experimental sites.

Sites	Primary Tillage	CC [1] Planting	CC Seeding Rate (kg ha^{-1})	CC Row Spacing (cm)	CC Rolling & Soybean Planting	Number of CC Growth Days	Soybean Seeding Rate (Seed ha^{-1})	Soybean Cultivar	Soybean Row Spacing (cm)
17-Arl. A1	16 August, 22 August, 2 September, 19 September, 2016	19 September (triticale), 26 September (rye) [2], 2016	201.75	19	26 May and 30 May (Rye)/8 June (Triticale), 2017	273 (rye)/62 (triticale)	555,986	Viking 0.1706	76.2
18-Arl. A2	29 September, 2 October, 2017	2 October, 2017	201.75	19	6 June (rye)/ 11 June (triticale), 2018	247 (rye)/252 (triticale)	555,986	Viking 0.1706	76.2
17-Frce B	9 July, 15 August, 15 September, 2016	23 September, 2016	200	16.5	16 May, 2017	235	605,000	ES Mentor	50
18-Frce C	15 July, 8 August, 21 August, 2017	22 August, 2017	200	12.5	29 May, 2018	280	535,000	Klaxon	50
18-Frce D	25 July, 10 August, 21 August, 24 August, 2017	25 August, 2017	200	12.5	18 May, 2018	266	600,000	ES Mentor	50
18-Frce E	18 July, 14 August, 25 August, 2017	29 August, 2017	200	16.5	25 May, 2018	269	535,000	ES Mentor	33

[1] Cover crop, [2] high amounts of precipitation during cover crop planting on 19 September prevented its completion on the same day. Wet conditions did not allow for the completion of cover crop planting until 26 September.

2.4. Statistical Analysis

Linear mixed models were used to evaluate the effect of rye and winter triticale on cover crop height and biomass, weed biomass (Date 1 and Date 2), soybean population, biomass at flowering and yield. "Cover crop species" was treated as a fixed effect. The six sites and eight plots per sites were treated as a random effect. The "site" factor refers to "location x year". The following model was used for analysis:

$$Yijk = Xi + Aj + Bk + Cjk + XEijk$$

where X is the fixed factor (cover crop species), A the first random effect (sites), B the second random effect (plots), C the interaction between both random effect factors, XE the error term, i a particular cover crop species, j a particular site (location × year) and k refer to a particular plot.

ANOVA per factor was also conducted for each site. Cover crop height, soybean height and yield met the assumptions for analysis of variance (ANOVA). Cover crop biomass, weed biomass at Date 1 and Date 2, as well as soybean density and biomass, were transformed as needed to meet the assumption for analysis of variance using square root transformation. We used the R software for every statistical analysis in R version 1.1.463 © RStudio, Inc, and more precisely the lme4 package for the linear mixed models [47]. Statistical significance of the results was evaluated at a p-value < 0.05 and treatment means were compared using Tukey's pairwise comparison.

3. Results

3.1. Climate

Rainfall accumulation during cover crop establishment was greater at all sites in the fall of 2016 compared to the fall 2017 (290 to 300 mm vs. less than 200 mm between September and November) (Figure 1). Both September and October were drier than average in southern France in 2017, while November and December were wetter. In April 2017 and May 2018, Arlington (site A), received more rain than average. The site received between 348.2 and 379.7 mm between April and June both in 2017 and in 2018 while the French sites only received between 173.4 and 216.4 mm over the same period. The greatest difference in rainfall accumulation between Arlington and the French sites was observed in the summer. While Arlington (site A) received 198.6 to 278.6 mm between July and August of 2017 and 2018 the French sites only received 72.9 to 129.7 mm over the same period (Figure 1).

In Arlington in 2017, monthly average temperatures were below 0 °C from November to April. The coldest months were December and January, with a minimum air temperature mean of −16.3 °C. In 2018, the temperature raised above 0 °C a month later than in 2017 (early May vs. early April) and the coldest months were January and February with monthly minimum air temperatures of −12.8 and −11.5 °C, respectively. At the French sites, both winters were milder than in Wisconsin and periods of freezing temperatures were rare. In 2017, at site B, January was the coldest month with −1.5 °C on average. The monthly average temperature was above 10 °C from March to the end of the growing season. In 2018, February was colder than December and January with one week of frost. Meteorological stations close to the C, D and E sites indicated a monthly minimum air temperature of −1.2 to −0.7 °C in February 2018 compared with 4.5 to 5.5 °C in January. Monthly average temperatures were above 10 °C at the French sites at the beginning of April 2018.

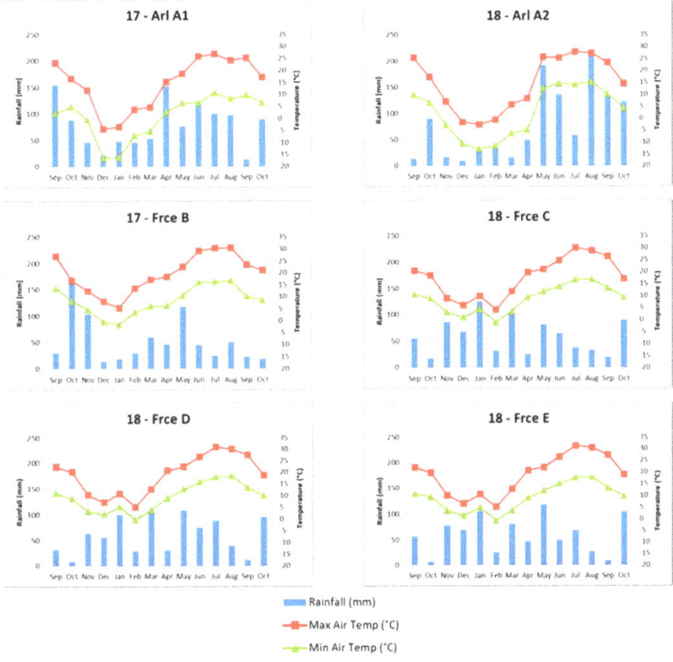

Figure 1. Monthly rainfall accumulation and average temperature for each of the six field locations over the 2016–2017 and 2017–2018 seasons.

3.2. Cover Crop Performance

Data from the six trials analyzed with linear mixed models did not show any significant difference in biomass production between cereal rye and triticale, with 6989 kg ha^{-1} and 7352 kg ha^{-1}, respectively (Figure 2). However, cereal rye grew significantly taller than triticale, 125 cm vs. 77 cm, respectively ($p < 0.001$).

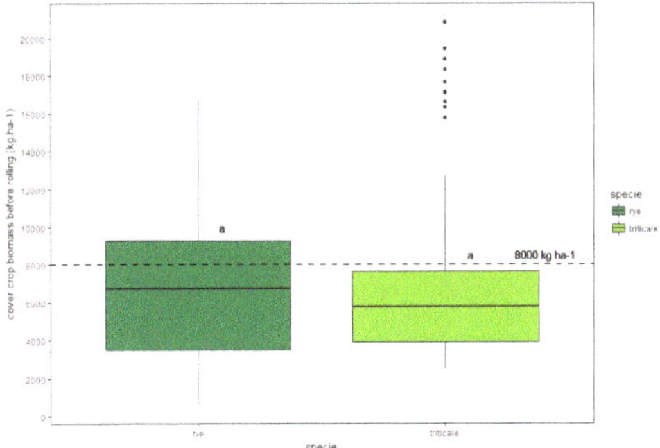

Figure 2. Mean weight of cereal rye and triticale biomass before cover crop rolling, averaged across all sites, 2017 and 2018. The linear mixed model did not indicate significant differences between cereal rye and triticale ($p > 0.05$, $n = 144$). Data presented in Figure 2 are means ± standard error. Each cover crop species (rye and triticale) followed by the same letter are not significantly different. The dotted line refers to a mean of the cover crop biomass values range reported in the scientific literature as a success factor to suppress weed until soybean harvest.

Cover crop biomass of both cereal rye and triticale was highly influenced by the pedoclimatic conditions (location x year) and varied from 2,963 kg ha^{-1} at the 18-Frce C site to 16,994 kg ha^{-1} at 17-Arl A1. Except for 18-Arl A2, the ANOVA performed per site showed a significant effect of cover crop species on cover crop biomass. However, the cover crop species producing the highest biomass differed between sites: Cereal rye for 17-Frce B and 18-Frce D sites and triticale for 17-Arl A1, 18-Frce C and 18-Frce E sites (Figure 3). In 2017, at Arlington, triticale produced significantly greater biomass than any other the site, resulting in the nine outlying data points seen in Figure 2.

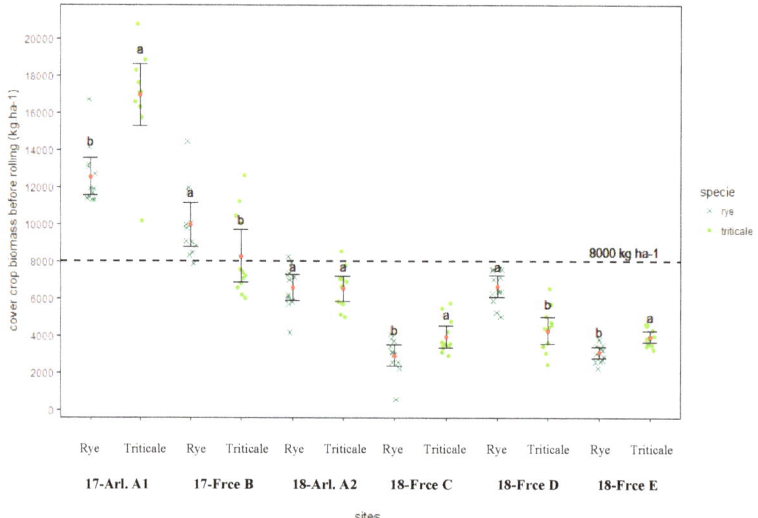

Figure 3. Cover crop biomass before termination per species and per site, 2017 and 2018. The letters represent the results of the ANOVAs per site ($p < 0.05$, $n = 24$). For each site, if the two species have the same letter their biomass is not significantly different. The dotted line refers to a mean of the cover crop biomass values range reported in the scientific literature as a success factor to suppress weed until soybean harvest.

3.3. Weed Biomass

In Wisconsin in 2018, the dominant weed species were Ladysthumb smartweed (*Polygonum persicaria* L.), lambsquarter (*Chenopodium album* L.), and foxtail (*Setaria pumila, Setaria viridis* and *Setaria faberi* L.). In France, the dominant weed species varied by location and year. At site B in 2017, common ragweed (*Ambrosia artemisiifolia* L.), heartsease (*Viola tricolor* L.) and Scarlet Pimpernel (*Anagallis arvensis*) dominated. In 2018, the weed population at site C was dominated by field bindweed (*Convolvulus arvensissuch* L.), all-seed (*Chenopodium polyspernum* L.) and round-leaved fluellin (*Kickxia spuria* L.) and at side D by yellow foxtail (*Setaria glauca* L.), switchgrass (*Panicum virgatum* L.) and persicaria (*Persicaria maculosa Gray* L.). Finally, in 2018 at site E, common ragweed was the main species along with annual bluegrass (*Poa annual* L.) and foxtail (*Setaria glauca* L.).

The linear mixed model determined that triticale provided poorer weed suppression compared to rye during the summer (Date 1), with 577 and 231 kg ha^{-1} of weed biomass for triticale and cereal rye, respectively (Table 3). A similar conclusion was shown in the fall (Date 2), with 1545 and 1178 kg ha^{-1} of weed biomass for triticale and rye, respectively. Weed dynamics between the summer and the fall did not differ significantly between rye and triticale. However, weed populations within the triticale cover crop tended to be higher than under cereal rye ($p = 0.09$) (Table 3). Indeed, data from the six trials indicated that the total weed biomass increased by an average of 945 kg ha^{-1} for the rye and 968 kg ha^{-1} for the triticale between Date 1 and Date 2.

Table 3. Summer and fall weed biomass from the two cover crop treatments over the six sites, 2017 and 2018. Weed biomass changes between the two dates, indicative of the degree of weed growth, are also reported.

Sites	Weed Biomass (kg ha^{-1}) Date 1 [1]		Weed Biomass (kg ha^{-1}) Date 2 [2]		Weed Biomass Change (kg ha^{-1}) Date 2–Date 1	
	Cereal Rye	Triticale	Cereal Rye	Triticale	Cereal Rye	Triticale
17-Arl. A1	83	148	287	274	204	127
18-Arl. A2	123	1214	55	869	−68	−345
17-Frce B	402	1119	319	857	−83	−262
18-Frce C	327	177	1245	1270	919	1093
18-Frce D	317	639	3479	4445	3163	3806
18-Frce E	134	163	1683	1554	1549	1391
Mean [3]	**231 a**	**577 b**	**1178 a**	**1545 b**	**947 a**	**968 a**
p-value [β]	<0.001		<0.001		0.09	
Significative Effect	***		***		.	

[1] Weed biomass was collected in July in France and in August at Arlington. [2] Weed biomass was collected in September. [3] Weed biomass mean from data of the six sites ($n = 6 \times 24$) (17-Arl. A1, 18-Arl. A2, 17-Frce B, 18-Frce C, 18-Frce D, 18 Frce E) are presented in bold in the table for each cover crop specie Cereal Rye and Triticale at the different dates of measurement (Date 1 and Date 2). [β] Linear mixed model, n = 144, Significance codes: 0 '***' 0.001 '**' 0.01 '*' 0.05 '.' 0.1 ' ' 1. Numbers in bold in the table followed by the different letters for Cereal Rye and Triticale within a similar date (Date 1, Date 2 or Date 2-Date 1) are significantly different.

The ANOVA per site showed that at the 18-Frce D site, the weed biomass was particularly high in the triticale plots with more than 4000 kg ha^{-1} (Table 3). Significant differences between triticale and cereal rye were also observed during the summer (Date 1) at the 18-Arl. site with 1214 kg ha^{-1} and 123 kg ha^{-1} of weed biomass, respectively. 18-Frce C is the only site where cereal rye resulted in poorer weed suppression compared to triticale. However, at that site, increased weed biomass between Date 1 and Date 2 in the triticale was observed as compared to rye (Table 3).

In France, except for site B where weed development was limited, an increase in weed biomass of more than 1000 kg ha^{-1} was observed between Date 1 and Date 2 for both cover crop species. At Arlington, weed biomass only increased by 204 kg ha^{-1} (cereal rye) and 127 kg ha^{-1} (triticale) between Date 1 and 2 in 2017 and decreased between the two measurements in 2018. Overall, the greatest increase in weed biomass between summer and fall was observed at 18-Frce C, D, and E (Table 3).

3.4. Soybean Performance

Using the linear mixed model, no differences were observed in soybean population between the two cover crop species, with an average of 309,020 plants ha^{-1} in the rye and 309,562 plants ha^{-1} in the triticale. At flowering, soybean biomass tended to be higher when planted into rye as compared to triticale, with 1,876 and 1,624 kg ha^{-1}, respectively ($p = 0.07$). Soybean height at flowering was also greater under rye cover crop (53 cm) compared to triticale (47 cm) ($p < 0.01$).

The linear mixed model indicated that the choice of cover crop species significantly affected soybean yields (Figure 4). Using cereal rye as opposed to triticale as a cover crop resulted in increased soybean yields of 0.1 to 1.3 t ha^{-1}. At the sites where triticale produced more biomass than rye, the yield gap between the two cover crop species was the lowest. At sites A1, C and E yields were only 0.2, 0.1 and 0.3 t ha^{-1} lower in the triticale (Table 4). The ANOVA per site also illustrated higher yields of soybeans grown with rye as compared to triticale, except for 18-Frce C and D where the variability within plots was high ($p > 0.05$) (Table 4). Independent of cover crop species, standard deviation varied from 0.8 to 0.9 t ha^{-1} at sites 18-Frce C, D, and E, it was 0.7 t ha^{-1} at 17-Arl. A1 and 18-Frce B and only 0.15 t ha−1 at 18-Arl. A2.

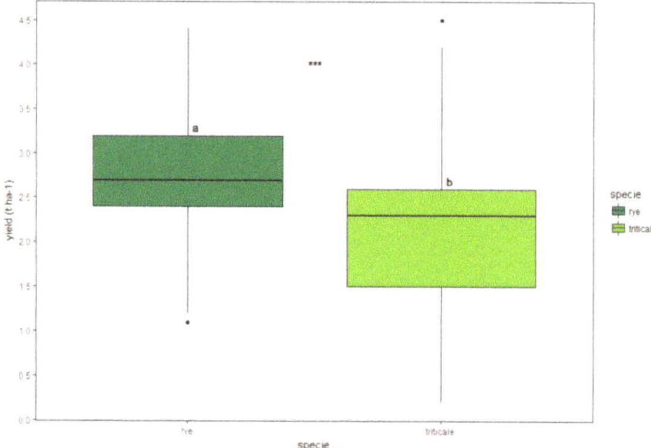

Figure 4. Soybean yields (t ha^{-1}) obtained from cereal rye and triticale cover crop treatments, averaged across all sites, 2017 and 2018, analyzed using linear mixed model (p < 0.001, n = 144). Boxplots followed by the same letters are not significantly different.

Table 4. Soybean yield (t ha^{-1}) averages per cover crop and per site, 2017 and 2018.

Sites	Soybean Yield (t ha^{-1})		Soybean Yield Losses on Triticale Cover Crop (t ha^{-1})
	Cereal Rye	Triticale	Cereal Rye-Triticale
17-Arl. A1	2.7 a	0.2 b	2.5
18-Arl. A2	3.5 a	2.2 b	1.3
17-Frce B	2.8 a	2.2 b	0.6
18-Frce C	2.7 a	2.6 a	0.1
18-Frce D	1.8 a	1.1 a	0.7
18-Frce E	2.6 a	2.3 b	0.3
Mean [1]	**2.7 a**	**2.2 b**	**0.5**
p-value (mean data) [β]	<0.001		-
Significative Effect	***		-

[β] Linear mixed model, n = 144, Significance codes: 0 '***' 0.001 '**' 0.01 '*' 0.05 '.' 0.1 ' ' 1. [1] Soybean yield mean from data of the six sites (n = 6 × 24) (17-Arl. A1, 18-Arl. A2, 17-Frce B, 18-Frce C, 18-Frce D, 18 Frce E) are presented in bold for: Cereal Rye, Triticale and Cereal Rye-Triticale cover crop. Numbers followed by the same letters under the same line in the table are not significantly different.

4. Discussion

4.1. Cover Crop Biomass Production

On average across all sites, despite the difference in cover crop height prior to rolling (i.e., rye taller than triticale), the cover crop biomass did not differ between the two species. However, ANOVA per site determined a significant effect of species on biomass production under certain pedoclimatic conditions, with the most productive species varying between years and locations (Figure 3). Among the six trials, cereal rye biomass before rolling varied from 2936 kg ha^{-1} (18-Frce C) to 12,588 kg ha^{-1} (17-Arl. A1) and triticale biomass ranged from 3977 kg ha^{-1} (18-Frce E) to 16,994 kg ha^{-1} (17-Arl. A1). Environment (soil and climate) was thus identified as a factor explaining part of the variability observed in the cover crop biomass. This was consistent with other findings in the scientific literature [38,45,48,49]. Smith et al. [37] also found contrasting results between years and locations in North Carolina on sandy and loamy sand soil characterized by a warm humid subtropical climate. For example, a decrease in rainfall accumulation in 2009 was correlated with lower rye biomass (4450 kg ha^{-1}) compared to the previous year (10,854 kg ha^{-1}).

As discussed by Smith et al. [37], improved cover crop management including fertilization, planting date, seeding rate, species, and cultivar choice is fundamental to successful cover crop establishment and biomass production. This "management x environment" effect was observed at site A, with more than 10,000 kg ha^{-1} of biomass for both species in 2017 and less than 7000 kg ha^{-1} in 2018. The first year, the cover crop was planted after alfalfa and manure was applied before planting. The second year, planting occurred after corn harvested for silage and did not receive manure, with temperatures reaching above 0 °C a month later than the previous year. The shorter period of cover crop biomass production at 18-Arl. A2, combined with both lower precipitation during cover crop establishment in the fall of 2017 and lower nitrogen availability, resulted in lower biomass.

A similar impact of both environment and management was observed in France, where in 2018, cover crop biomass was lower than 4000 kg ha^{-1} for every species at every site except cereal rye on site D. The 2017–2018 growing season was characterized by below-average rainfall during cover crop establishment which affected cover crop emergence followed by above average rainfall in the winter which led to reduced tillering. With a fine loam clay soil type, the 18-Frce C was the most affected by the wet conditions. The water did not readily infiltrate through the soil, leading to cover crop stand losses (e.g., 2963 kg ha^{-1} of cereal rye biomass). A week of frost in February after mild January temperatures which had brought the cover crops out of dormancy also negatively impacted cover crop development in France in 2018. At site D, the earlier planting date (25 August), nitrogen credit from the preceding alfalfa crop, and mild fall temperatures (above 10 °C until November) led to rapid cover crop development before winter. The cover crop was thus at more sensitive stage than at other locations during the period of frost in February, which affected its biomass production potential. The significant difference in cover crop biomass between rye and triticale at site D in 2018 (6668 and 4314 kg ha^{-1}, respectively) was likely explained by the superior winter hardiness of rye compared to triticale (Figure 3).

Cover crop planting and termination timing have often been observed to play a key role in cover crop biomass production, explaining part of the variability between sites [50,51]. Mirsky et al. [33] discussed the increase of cover crop biomass production in May in mid-Atlantic region of US following earlier cover crop planting by comparing six planting dates across 10 day intervals under high annual precipitation condition evenly distributed (760–1012 mm) and silt loam soil. Delayed cover crop termination is critical to both improve cover crop termination and increase biomass production. Depending on specific annual conditions, cereal biomass can increase by 200 kg ha^{-1} per day after the stem elongation stage (i.e., after the 39 Zadok stage) [17]. In our study, planting dates varied from mid-August to early October and termination dates from mid-May to mid-June (Table 2).

One strategy to increase the resilience of the CCBRT may include the use of cover crop species mixtures. As suggested by Liebert et al. [39] in New York, mixing tall species such as cereal rye with species that are shorter with wider leaves such as triticale or barley can optimize early soil shading and hasten canopy closure. This strategy could improve cover crop establishment and early spring weed control as well as increase the probability of achieving adequate cover crop biomass at rolling under challenging conditions (e.g., soil type heterogeneity, drought, excess of water, etc.). The main drawback of using species mixtures is the lack of synchronization of anthesis of the different cultivars, which would need to be assessed for successful implementation of a roll-crimp system. Cover crop termination of cereal using a roller-crimper has been shown to be most effective when done between anthesis and early dough stage (Zadoks growth stage 61 to 85), with termination increasingly effective as the cereal matures to the soft dough stage [16,17,40,48,51].

The different cover crop cultivars used in the study particularly between the Upper Midwest and Southern France trials have to be considered as cultivar might impact the potential of cover crop biomass production, the cover crop sensitivity to cold temperatures and change climate as well as the cover crop flowering period [40,52]. In North Carolina, Wells et al. [40] observed higher cereal rye biomass (>9000 kg ha^{-1}) and greater cover crop control (100%) using earlier-flowering cultivar compared with late-flowering cultivar where cover crop biomass was inferior to 9000 kg ha^{-1} and the cover crop control effectiveness was inferior to 65%. Thus, despite the cover crop biomass effect, ability to provide an adequate cover crop termination also might influence the weed pressure as well as soybean yield. To address the cultivar effect, interest in breed early-flowering fall rye is growing in North America to a specific adaptation for organic CCBRT as observed in Canada with the "CETAB + HÂTIF" cultivar [53,54].

4.2. Weed Biomass

As observed by Liebert and Ryan [55] and Ryan et al. [56] in humid continental climate and silt loam soil, results showed that when adequate biomass is produced prior to termination, the cover crop can significantly limit weed development. A sufficient amount of cover crop biomass remaining on the soil surface can reduce weed development by acting as a physical barrier, competing with weeds for nutrients, light, and water, and releasing allelopathic compounds [20,57,58]. Previous research has concluded that cover crop biomass should reach from 6000 to 10,000 kg ha^{-1} before termination to ensure adequate weed control until cash crop harvest, with more reliable control at biomass rates closer to 8000 kg ha^{-1} [17,18,37]. In our study, high levels of cover crop biomass (>8000 kg ha^{-1}) were reached at two sites: 17-Arl. A1 and 17-Frce B (Figure 3). At these sites, weed biomass increased by 127 to 204 kg ha^{-1} between Date 1 and Date 2, while at the other southern French sites, the weed biomass increased by more than 1000 kg ha^{-1} within the same timeframe. In Wisconsin, within the 18-Arl. A2 conditions, cover crop biomasses averaged 6615 and 6548 kg ha^{-1}, which although on the lower end of the anticipated acceptable range suppressed weed establishment throughout soybean season.

While species did not significantly differ in their biomass produced in our multi-site comparative study, cover crop species did differ in their weed suppression. Indeed, results showed that compared to triticale, cereal rye more effectively suppressed weeds through the entire soybean growing season. These results were consistent with previous organic CCBRT studies conducted in soybean or corn production systems. These studies also found that cereal rye used as a cover crop in CCBRT systems provided better weed control than other winter cereals or mixes of winter cereals and legume cover crops (e.g., winter wheat and winter pea, winter wheat, hairy vetch) [18,59,60]. In Iowa, located within the same cold temperate climate as Wisconsin, Delate et al. [18] observed lower weed pressure (broadleaf species) on silty clay loam soil with a cover crop mixture including cereal rye and hairy vetch compared to a mix of wheat and winter pea, with weed populations at the beginning of June of 2.2 plant m-^{2} and 6.5 plant m-^{2}, respectively. According to numerous researchers, the greater allelopathic effect of cereal rye may explain the greater weed control observed [61–63]. While few published studies directly address this phenomenon, within these systems where the cover crop remains on the soil surface, the release of allelopathic compounds could be delayed providing greater season-long effects [42].

Several studies have compared cereal rye and triticale as cover crops in organic CCBRT soybean production system. In conventional systems in Ontario, Canada, Moore et al. [60] indicated that cereal rye provided better control of redroot pigweed (*Amaranthus retroflexus* L.) than triticale and wheat. In organic systems, Silva [45] did not find any difference in weed biomass using cereal rye, triticale or barley as cover crop neither before cover crop termination nor 12 weeks after cover crop termination in Wisconsin in 2010 and 2011. These results contrast with our study, and the difference could be explained by the lower variability in cover crop biomass observed by Silva [45] in 2010 and 2011. In our study, the cover crop biomass was particularly low at three out of six sites (e.g., biomass less than 5000 kg ha^{-1}) while Silva [45] obtained more than 10,000 kg ha^{-1} of cereal rye, triticale and barley cover crop in 2010 and 2011 (with the exception of triticale in 2011 with 6380 kg ha^{-1}). When cover crop biomass is lower than 8000 kg ha−1, according to Teasdale and Mohler [64] and Smith et al. [37] a difference of 1000 to 2000 kg ha^{-1} of rye biomass between cover crops may explain the success or failure of a CCBRT system. The broad range of pedoclimatic conditions encountered in our study did not allow for the confirmation of this hypothesis, but greater weed growth was observed between summer and fall when the cover crop biomass was less than 6000 kg ha^{-1}. At 18-Frce C, D and E sites, where the cover crop biomass was less than 5000 kg ha^{-1}, weed biomass increased between Date 1 and Date 2 was high (918 to 3162 kg ha^{-1}). Conversely, at the 17-Arl. A1, 18-Arl. A2 and 17-Frce B sites where cover crop biomass was greater than 6000 kg ha^{-1} before rolling, the weed biomass remained stable or increased only slightly.

Despite of cover crop biomass and allelopathic effect, others factors related to the species characteristics also might influence weed management such as potential of tiller number production ensuring soil covering, leaf area, vegetative/reproductive ratio, decay dynamic of cover crop on soil surface (i.e., C/N ratio) and root growth [65]. These remain poorly documented in the literature, but recent promising paper promote the interest in species mixtures which can provide benefits for weed management and hasten canopy cover before cover crop rolling [39]. For instance, cereal rye combined with other species characterized by shorter height and wider leaves such as triticale or barley could increase light interception and shading.

4.3. Soybean Yield

Soybean emergence did not differ between cover crop species or between sites. On average, with a seeding rate between 535,000 and 605,000 seed ha^{-1}, resulting stands only reached 309,291 plant ha^{-1}. In Iowa, with the same seeding rate, Delate et al. [18] also observed poor emergence with a final stand count of 324,000 plant ha^{-1}. According to Wallace et al., in order to improve soybean emergence, major improvements to no-till planters must occur. To ensure appropriate seed-to-soil contact, no-till planters must slice through a thick cover crop mulch prior to opening and closing the planting furrow, as a poor seeding environment can result in poor soybean emergence and thereby affect soybean yields [21,66].

While not impacting soybean emergence, cover crop species treatments differed in their subsequent soybean yields. The cereal rye treatments resulted in significantly greater soybean yields as compared to using triticale, with 2.7 and 2.2 t ha^{-1}, respectively. In Pennsylvania, US, with humid continental climate and Southwest Germany, Europe, with moderately continental climate, Wallace et al. [21] and Weber et al. [66] also compared cereal rye with other cereal species as cover crops (i.e., barley (*Hordeum vulgare* L.), but did not observe any difference in soybean yield. To explain these results, the authors concluded that depending on the conditions, barley can produce adequate weed control due to quicker canopy closure and wider leaf blades compared to rye. Thus, combining rye with barley can result in similar weed control to the rye cover crop alone, thus leading to similar soybean yields. In our study, when the triticale produced equivalent or higher levels of biomass than rye, it provided equivalent weed suppression between Date 1 and Date 2 than rye, thereby limiting the yield loss observed on triticale compared with rye cover crop (17-Arl. A1, 17-Frce B, 18-Arl. A2).

Independent of the cover crop species, our results showed that the variability within plots (as measured by standard deviation) is increased in situations where the cover crop biomass is low (i.e., < 6000 kg ha^{-1}). The improved weed control provided by cover crop biomass in excess of 6000 kg ha^{-1} (i.e., 17-Arl. A1, 18-Arl. A2, and 17-Frce B) reduces water and nutrient competition between weeds and soybean plants, resulting in both more consistent and higher yields. However, cover crop biomass does not appear as the main factor explaining resultant soybean yields: While the average cover crop biomass did not differ among sites, the highest yields were obtained when planting soybean into rye. Additionally, soybean emergence does not seem to explain yield differences. Weed species and growth over the season, influenced by both (i) initial cover crop biomass before rolling and (ii) cover crop species, appeared as a driving factor impacting yields in our study. The allelopathic effect of rye likely influenced weed emergence as well, explaining the higher yields obtained compared to soybean planted into triticale.

5. Conclusions

This study illustrated the impact of the pedoclimatic condition on cover crop biomass produced using CCBRT systems, which subsequently impacted weed species and biomass dynamics throughout the soybean growing season. Despite location and year effect, choice of cover crop species remains a fundamental decision for adequate weed suppression and sustainable soybean yields. Specifically, the results show that cereal rye remains the best candidate for successful organic CCBRT soybean production. The allelopathic effect of cereal rye likely suppresses weed seed germination to a greater degree than what is achieved by other annual cereal grain species. In addition, cereal rye is more winter hardy and reaches anthesis earlier than triticale, benefiting both biomass accumulation and timely planting with the roller-crimper. Our results provide further confirmation that sufficient cover crop biomass is crucial to suppress weeds throughout the cash crop production season. However, depending on location and year (e.g., dry, wet, degree and length of time below freezing), failures in cover crop establishment and/or poor development may be encountered with cereal rye and can lead to significant yield losses. On a farm-scale level, moving beyond evaluating cover crop decisions solely on agronomic performance, economic and practical considerations may impact farmer's choice. For example, cereal rye seed can be more expensive and difficult to access in some regions, such as southern France, as compared to other cereal grain species. With the high seeding rates needed in CCBRT systems, seed cost is a critical factor in the net profitability of the system. Mixing cereal rye with another high biomass cereal species such as triticale may allow for beneficial aspects of both species, including maximizing soil coverage among a variety of soil and climate conditions while reducing seed costs. In addition to the multi-tactic strategies previously highlighted to optimize cover crop biomass production (e.g., planting date, fertilization, irrigation, etc.), additional cover crop varieties and species mixes should be considered for further research (e.g., forage rye, forest rye, etc.). More broadly, in a changing climate, future CCBRT research should focus on flexible decision-support tools based on multi-tactic cover crop management to assist farmers in making the best decisions to ensure cover crop performance and weed management throughout the cash crop growing season.

Author Contributions: L.V.-C. reviewed the literature and wrote the initial draft of the paper with the assistance from L.V. Then L.V-C, L.V, E.S. and J.P., contributed to revising the manuscript.

Funding: The financial support for this project provided by transnational funding bodies, including partners of the FP7 ERA-net project, CORE Organic Plus, and cofunds from the European Commission, the TERRA ISARA foundation, the French Water Agency 'Rhône Méditerranée Corse'. The USA portion of the work was partially funded by the Ceres Trust and by a USDA ARS Cooperative Agreement (58-5090-7-072). The text in this paper is the sole responsibility of the authors and does not necessarily reflect the views of the national and European funding bodies that financed this project.

Acknowledgments: We acknowledge V. Payet for his assistance on the statistical analysis. We would like to thank the technical teams from the University of Wisconsin-Madison and ISARA. We would like to give a special thanks to the organic farmers who carried out the French trials.

Conflicts of Interest: The authors declare no conflict of interest.

References

1. Willer, H.; Lernoud, J. *The World of Organic Agriculture—Statistics and Emerging Trends*; FiBL and IFOAM: Frick, Switzerland, 2019; p. 353.
2. USDA Economics, Statistics and Market Information System. Available online: https://usda.library.cornell.edu/ (accessed on 27 May 2019).
3. U.S. Department of Agriculture, Economic Research Service. Organic Market Overview. Available online: https://www.ars.usda.gov/research/publications/publication/?seqNo115=178152 (accessed on 25 February 2018).
4. Agence Bio La Bio dans l'Union Européenne. Available online: https://www.agencebio.org/wp-content/uploads/2018/10/Carnet_UE_2017.pdf (accessed on 27 May 2019).
5. Delate, K.; DeWitt, J. Building a farmer-centered land grant university organic agriculture program: A Midwestern partnership. *Renew. Agric. Food Syst.* **2004**, *19*, 80–91. [CrossRef]
6. Riemens, M.M.; Groeneveld, R.M.W.; Lotz, L.A.P.; Kropff, M.J. Effects of three management strategies on the seedbank, emergence and the need for hand weeding in an organic arable cropping system. *Weed Res.* **2007**, *47*, 442–451. [CrossRef]

7.   Watson, C.A.; Atkinson, D.; Gosling, P.; Jackson, L.R.; Rayns, F.W. Managing soil fertility in organic farming systems. *Soil Use Manag.* **2002**, *18*, 239–247. [CrossRef]

8.   Baker, B.P.; Benbrook, C.M.; Groth, E.; Lutz Benbrook, K. Pesticide residues in conventional, integrated pest management (IPM)-grown and organic foods: Insights from three US data sets. *Food Addit. Contam.* **2002**, *19*, 427–446. [CrossRef] [PubMed]

9.   Hole, D.G.; Perkins, A.J.; Wilson, J.D.; Alexander, I.H.; Grice, P.V.; Evans, A.D. Does organic farming benefit biodiversity? *Biol. Conserv.* **2005**, *122*, 113–130. [CrossRef]

10. Anderson, R.L. A Multi-Tactic Approach to Manage Weed Population Dynamics in Crop Rotations. *Agron. J.* **2005**, *97*, 1579–1583. [CrossRef]

11. Bàrberi, P. Weed management in organic agriculture: Are we addressing the right issues? *Weed Res.* **2002**, *42*, 177–193. [CrossRef]

12. Maeder, P.; Fliessbach, A.; Dubois, D.; Gunst, L.; Fried, P.; Niggli, U. Soil Fertility and Biodiversity in Organic Farming. *Science* **2002**, *296*, 1694–1697. [CrossRef] [PubMed]

13. FAO Economie de l'agriculture de Conservation. Available online: http://www.fao.org/docrep/005/y2781f/y2781f03.htm (accessed on 8 March 2015).

14. Holland, J.M. The environmental consequences of adopting conservation tillage in Europe: Reviewing the evidence. *Agric. Ecosyst. Environ.* **2004**, *103*, 1–25. [CrossRef]

15. Berner, A.; Hildermann, I.; Fließbach, A.; Pfiffner, L.; Niggli, U.; Mäder, P. Crop yield and soil fertility response to reduced tillage under organic management. *Soil Tillage Res.* **2008**, *101*, 89–96. [CrossRef]

16. Moyer, J. *Organic No-Till Farming. Advancing No-Till Agriculture. Crops, Soil, Equipment*; Acres U.S.A.: Austin, TX, USA, 2011; ISBN 978–1-60173-017-6.

17. Mirsky, S.B.; Ryan, M.R.; Curran, W.S.; Teasdale, J.R.; Maul, J.; Spargo, J.T.; Moyer, J.; Grantham, A.M.; Weber, D.; Way, T.R.; et al. Conservation tillage issues: Cover crop-based organic rotational no-till grain production in the mid-Atlantic region, USA. *Renew. Agric. Food Syst.* **2012**, *27*, 31–40. [CrossRef]

18. Delate, K.; Cwach, D.; Fiscus, M. Evaluation of an Organic No-Till System for Organic Corn and Soybean Production–Agronomy Farm Trial, 2011. In *Organic Ag Program Webpage*; Iowa State University: Ames, IA, USA, 2012.

19. Silva, E.; Delate, K. A Decade of Progress in Organic Cover Crop-Based Reduced Tillage Practices in the Upper Midwestern USA. *Agriculture* **2017**, *7*, 44. [CrossRef]

20. Teasdale, J.R.; Coffman, C.B.; Mangum, R.W. Potential long-term benefits of no-tillage and organic cropping systems for grain production and soil improvement. *Agron. J.* **2007**, *99*, 1297–1305. [CrossRef]

21. Wallace, J.; Williams, A.; Liebert, J.; Ackroyd, V.; Vann, R.; Curran, W.; Keene, C.; VanGessel, M.; Ryan, M.; Mirsky, S. Cover Crop-Based, Organic Rotational No-Till Corn and Soybean Production Systems in the Mid-Atlantic United States. *Agriculture* **2017**, *7*, 34. [CrossRef]

22. Blanchart, E.; Bernoux, M.; Sarda, X.; Siqueira Neto, M.; Cerri, C.C.; Piccolo, M.; Douzet, J.-M.; Scopel, E.; Feller, C. Effect of direct seeding mulch-based systems on soil carbon storage and macrofauna in Central Brazil. *Agric. Conspec. Sci. ACS* **2007**, *72*, 81–87.

23. Wayman, S.; Cogger, C.; Benedict, C.; Burke, I.; Collins, D.; Bary, A. The influence of cover crop variety, termination timing and termination method on mulch, weed cover and soil nitrate in reduced-tillage organic systems. *Renew. Agric. Food Syst.* **2015**, *30*, 450–460. [CrossRef]

24. Kunz, C.; Sturm, D.J.; Varnholt, D.; Walker, F.; Gerhards, R. Allelopathic effects and weed suppressive ability of cover crops. *Plant Soil Environ.* **2016**, *62*, 60–66.

25. Dhima, K.V.; Vasilakoglou, I.B.; Eleftherohorinos, I.G.; Lithourgidis, A.S. Allelopathic Potential of Winter Cereals and Their Cover Crop Mulch Effect on Grass Weed Suppression and Corn Development. *Crop Sci.* **2006**, *46*, 345–352. [CrossRef]

26. Jabran, K.; Mahajan, G.; Sardana, V.; Chauhan, B.S. Allelopathy for weed control in agricultural systems. *Crop Prot.* **2015**, *72*, 57–65. [CrossRef]

27. Mäder, P.; Berner, A. Development of reduced tillage systems in organic farming in Europe. *Renew. Agric. Food Syst.* **2012**, *27*, 7–11. [CrossRef]

28. Vincent-Caboud, L.; Peigné, J.; Casagrande, M.; Silva, E.M. Overview of Organic Cover Crop-Based No-Tillage Technique in Europe: Farmers' Practices and Research Challenges. *Agriculture* **2017**, *7*, 42. [CrossRef]

29. Zikeli, S.; Gruber, S. Reduced Tillage and No-Till in Organic Farming Systems, Germany—Status Quo, Potentials and Challenges. *Agriculture* **2017**, *7*, 35. [CrossRef]

30. Lee, J.A.; Gill, T.E. Multiple causes of wind erosion in the Dust Bowl. *Aeolian Res.* **2015**, *19*, 15–36. [CrossRef]
31. Friedrich, T.; Kassam, A.; Corsi, S.; Jat, R.A.; Sahrawat, K.L.; Kassam, A.H. Conservation agriculture in Europe. In *Conservation Agriculture: Global Prospects and Challenges*; CABI: Oxfordshire, UK; Boston, MA, USA, 2014; pp. 127–170.
32. Casagrande, M.; Peigné, J.; Payet, V.; Mäder, P.; Sans, F.X.; Blanco-Moreno, J.M.; Antichi, D.; Bàrberi, P.; Beeckman, A.; Bigongiali, F.; et al. Organic farmers' motivations and challenges for adopting conservation agriculture in Europe. *Org. Agric.* **2015**, *6*, 281–295. [CrossRef]
33. Mirsky, S.B.; Ryan, M.R.; Teasdale, J.R.; Curran, W.S.; Reberg-Horton, C.S.; Spargo, J.T.; Wells, M.S.; Keene, C.L.; Moyer, J.W. Overcoming Weed Management Challenges in Cover Crop–Based Organic Rotational No-Till Soybean Production in the Eastern United States. *Weed Technol.* **2013**, *27*, 193–203. [CrossRef]
34. Kornecki, T.S.; Price, A.J. Effects of different roller/crimper designs and rolling speed on rye cover crop termination and seed cotton yield in a no-till system. *J. Cotton Sci.* **2010**, *14*, 212–220.
35. Reberg-Horton, S.C.; Grossman, J.M.; Kornecki, T.S.; Meijer, A.D.; Price, A.J.; Place, G.T.; Webster, T.M. Utilizing cover crop mulches to reduce tillage in organic systems in the southeastern USA. *Renew. Agric. Food Syst.* **2012**, *27*, 41–48. [CrossRef]
36. Keene, C.L.; Curran, W.S.; Wallace, J.M.; Ryan, M.R.; Mirsky, S.B.; VanGessel, M.J.; Barbercheck, M.E. Cover Crop Termination Timing is Critical in Organic Rotational No-Till Systems. *Agron. J.* **2017**, *109*, 272–282. [CrossRef]
37. Smith, A.N.; Reberg-Horton, S.C.; Place, G.T.; Meijer, A.D.; Arellano, C.; Mueller, J.P. Rolled Rye Mulch for Weed Suppression in Organic No-Tillage Soybeans. *Weed Sci.* **2011**, *59*, 224–231. [CrossRef]
38. Clark, K.M.; Boardman, D.L.; Staples, J.S.; Easterby, S.; Reinbott, T.M.; Kremer, R.J.; Kitchen, N.R.; Veum, K.S. Crop Yield and Soil Organic Carbon in Conventional and No-till Organic Systems on a Claypan Soil. *Agron. J.* **2017**, *109*, 588–599. [CrossRef]
39. Liebert, J.A.; DiTommaso, A.; Ryan, M.R. Rolled Mixtures of Barley and Cereal Rye for Weed Suppression in Cover Crop–based Organic No-Till Planted Soybean. *Weed Sci.* **2017**, *65*, 426–439. [CrossRef]
40. Wells, M.S.; Brinton, C.M.; Reberg-Horton, S.C. Weed suppression and soybean yield in a no-till cover-crop mulched system as influenced by six rye cultivars. *Renew. Agric. Food Syst.* **2015**, *31*, 429–440. [CrossRef]
41. Tabaglio, V.; Gavazzi, C.; Schulz, M.; Marocco, A. Alternative weed control using the allelopathic effect of natural benzoxazinoids from rye mulch. *Agron. Sustain. Dev.* **2008**, *28*, 397–401. [CrossRef]
42. Schulz, M.; Marocco, A.; Tabaglio, V.; Macias, F.A.; Molinillo, J.M.G. Benzoxazinoids in Rye Allelopathy—From Discovery to Application in Sustainable Weed Control and Organic Farming. *J. Chem. Ecol.* **2013**, *39*, 154–174. [CrossRef]
43. Bernstein, E.R.; Posner, J.L.; Stoltenberg, D.E.; Hedtcke, J.L. Organically Managed No-Tillage Rye–Soybean Systems: Agronomic, Economic, and Environmental Assessment. *Agron. J.* **2011**, *103*, 1169. [CrossRef]
44. De Bruin, J.L.; Porter, P.M.; Jordan, N.R. Use of a Rye Cover Crop following Corn in Rotation with Soybean in the Upper Midwest. *Agron. J.* **2005**, *97*, 587–598. [CrossRef]
45. Silva, E.M. Screening Five Fall-Sown Cover Crops for Use in Organic No-Till Crop Production in the Upper Midwest. *Agroecol. Sustain. Food Syst.* **2014**, *38*, 748–763. [CrossRef]
46. Peigné, J.; Lefevre, V.; Vian, J.F.; Fleury, P. *Conservation Agriculture in Organic Farming: Experiences, Challenges and Opportunities in Europe*; Springer International Publishing: Cham, Switzerland, 2015; ISBN 978-3-319-11619-8.
47. Bates, D.; Mächler, M.; Bolker, B.; Walker, S. Fitting Linear Mixed-Effects Models using lme4. *arXiv* **2014**, arXiv:1406.5823.
48. Ashford, D.L.; Reeves, D.W. Use of a mechanical roller-crimper as an alternative kill method for cover crops. *Am. J. Altern. Agric.* **2003**, *18*, 37–45. [CrossRef]
49. Crowley, K.A.; Van Es, H.M.; Gómez, M.I.; Ryan, M.R. Trade-Offs in Cereal Rye Management Strategies Prior to Organically Managed Soybean. *Agron. J.* **2018**, *110*, 1492–1504. [CrossRef]
50. Mischler, R.A.; Curran, W.S.; Duiker, S.W.; Hyde, J.A. Use of a Rolled-rye Cover Crop for Weed Suppression in No-Till Soybeans. *Weed Technol.* **2010**, *24*, 253–261. [CrossRef]
51. Mirsky, S.B.; Curran, W.S.; Mortenseny, D.M.; Ryan, M.R.; Shumway, D.L. Timing of Cover-Crop Management Effects on Weed Suppression in No-Till Planted Soybean using a Roller-Crimper. *Weed Sci.* **2011**, *59*, 380–389. [CrossRef]

52. Neu, K.; Nair, A. *Effect of Planting Date and Cultivar on Cereal Rye Development and Termination for Organic No-Till Production Systems*; Iowa State University, Digital Repository: Ames, IA, USA, 2017.

53. Halde, C.; Gagné, S.; Charles, A.; Lawley, Y. Organic No-Till Systems in Eastern Canada: A Review. *Agriculture* **2017**, *7*, 36. [CrossRef]

54. La France, D.; Comeau, A.; Duval, J. *Développement d'un Seigle Adapté au rôle de Couvre-sol Pour le Semis Direct Sans Herbicide*; Final Report; CETAB+: Quebec City, QC, Canada, 2016; p. 36.

55. Liebert, J.A.; Ryan, M.R. High planting rates improve weed suppression, yield, and profitability in organically-managed no-till planted soybean. *Weed Technol.* **2017**, *31*, 536–549. [CrossRef]

56. Ryan, M.R.; Curran, W.S.; Grantham, A.M.; Hunsberger, L.K.; Mirsky, S.B.; Mortensen, D.A.; Nord, E.A.; Wilson, D.O. Effects of Seeding Rate and Poultry Litter on Weed Suppression from a Rolled Cereal Rye Cover Crop. *Weed Sci.* **2011**, *59*, 438–444. [CrossRef]

57. Carr, P.; Gramig, G.; Liebig, M. Impacts of Organic Zero Tillage Systems on Crops, Weeds, and Soil Quality. *Sustainability* **2013**, *5*, 3172–3201. [CrossRef]

58. Bernstein, E.R.; Stoltenberg, D.E.; Posner, J.L.; Hedtcke, J.L. Weed Community Dynamics and Suppression in Tilled and No-Tillage Transitional Organic Winter Rye–Soybean Systems. *Weed Sci.* **2014**, *62*, 125–137. [CrossRef]

59. Parr, M.; Grossman, J.M.; Reberg-Horton, S.C.; Brinton, C.; Crozier, C. Nitrogen Delivery from Legume Cover Crops in No-Till Organic Corn Production. *Agron. J.* **2011**, *103*, 1578. [CrossRef]

60. Moore, M.J.; Gillespie, T.J.; Swanton, C.J. Effect of Cover Crop Mulches on Weed Emergence, Weed Biomass, and Soybean (Glycine max) Development. *Weed Technol.* **1994**, *8*, 512–518. [CrossRef]

61. Weston, L.A. Utilization of Allelopathy for Weed Management in Agroecosystems. *Agron. J.* **1996**, *88*, 860–866. [CrossRef]

62. Reberg-Horton, S.C.; Burton, J.D.; Danehower, D.A.; Ma, G.; Monks, D.W.; Murphy, J.P.; Ranells, N.N.; Williamson, J.D.; Creamer, N.G. changes over time in the allelochemical content of ten cultivars of rye (*Secale cereale* L.). *J. Chem. Ecol.* **2005**, *31*, 179–193. [CrossRef] [PubMed]

63. Altieri, M.A.; Lana, M.A.; Bittencourt, H.V.; Kieling, A.S.; Comin, J.J.; Lovato, P.E. Enhancing Crop Productivity via Weed Suppression in Organic No-Till Cropping Systems in Santa Catarina, Brazil. *J. Sustain. Agric.* **2011**, *35*, 855–869. [CrossRef]

64. Teasdale, J.R.; Mohler, C.L. The quantitative relationship between weed emergence and the physical properties of mulches. *Weed Sci.* **2000**, *48*, 385–393. [CrossRef]

65. Halde, C.; Entz, M.H. Plant species and mulch application rate affected decomposition of cover crop mulches used in organic rotational no-till systems. *Can. J. Plant Sci.* **2016**, *96*, 59–71. [CrossRef]

66. Weber, J.; Kunz, C.; Peteinatos, G.; Zikeli, S.; Gerhards, R. Weed Control Using Conventional Tillage, Reduced Tillage, No-Tillage, and Cover Crops in Organic Soybean. *Agriculture* **2017**, *7*, 43. [CrossRef]

Article

Greenhouse Gas Emissions from Soil Cultivated with Vegetables in Crop Rotation under Integrated, Organic and Organic Conservation Management in a Mediterranean Environment

Simona Bosco [1,*,†]**, Iride Volpi** [1,†]**, Daniele Antichi** [2]**, Giorgio Ragaglini** [1] **and Christian Frasconi** [2]

[1]   Institute of Life Sciences, Scuola Superiore Sant'Anna, Via Santa Cecilia, 3, 56127 Pisa, Italy
[2]   Department of Agriculture, Food and Environment, University of Pisa, Via del Borghetto, 80, 56124 Pisa, Italy
*   Correspondence: s.bosco@santannapisa.it; Tel.: +39-050-883512
†   These authors contributed equally.

Received: 2 July 2019; Accepted: 11 August 2019; Published: 13 August 2019

Abstract: A combination of organic and conservation approaches have not been widely tested, neither considering agronomic implications nor the impacts on the environment. Focussing on the effect of agricultural practices on greenhouse gas (GHG) emissions from soil, the hypothesis of this research is that the organic conservation system (ORG+) may reduce emissions of N_2O, CH_4 and CO_2 from soil, compared to an integrated farming system (INT) and an organic (ORG) system in a two-year irrigated vegetable crop rotation set up in 2014, in a Mediterranean environment. The crop rotation included: Savoy cabbage (*Brassica oleracea* var. sabauda L. cv. Famosa), spring lettuce (*Lactuca sativa* L. cv. Justine), fennel (*Foeniculum vulgare* Mill. cv. Montebianco) and summer lettuce (*L. sativa* cv. Ballerina). Fluxes from soil of N_2O, CH_4 and CO_2 were measured from October 2014 to July 2016 with the flow-through non-steady state chamber technique using a mobile instrument equipped with high precision analysers. Both cumulative and daily N_2O emissions were mainly lower in ORG+ than in INT and ORG. All the cropping systems acted as a sink of CH_4, with no significant differences among treatments. The ORG and ORG+ systems accounted for higher cumulative and daily CO_2 emissions than INT, maybe due to the stimulating effect on soil respiration of organic material (fertilizers/plant biomass) supplied in ORG and ORG+. Overall, the integration of conservation and organic agriculture showed a tendency for higher CO_2 emissions and lower N_2O emissions than the other treatments, without any clear results on its potential for mitigating GHG emissions from soil.

Keywords: no-till; cover crops; green manure; organic fertilizers; carbon dioxide; methane; nitrous oxide

1. Introduction

In the last years, different concepts of sustainable agriculture have been proposed to increase food production while minimizing environmental impacts and maintaining economic sustainability.

Among them integrated farming (INT) has been promoted as a compromise between the reduction of the negative impacts of agricultural production on the environment and the economic sustainability of farms, and it has been described as a "third way" between conventional and organic agriculture [1,2].

Beside INT, other more challenging agricultural models has been proposed, such as the one proposing the integration between conservation agriculture and organic agriculture [3,4].

Conservation agriculture has been identified as: (i) A strategy for climate change adaptation, because it may increase soil organic matter improving resilience to extreme events, and (ii) for greenhouse gas (GHG) emissions mitigation thanks to the potentially improved carbon sequestration [5–7]. However, weed control remains one of the major issues under conservative tillage, thus the use of synthetic herbicides is required [8].

Organic farming is one of the main forms of agriculture that aims to balance the demands of food safety with environmental sustainability. Although the adoption of conservative tillage is also recommended in organic farming [9], several practices normally adopted in organic systems, and above all in vegetable production, imply frequent soil disturbance. Indeed, weed control is usually carried out through mechanical operations, including also ploughing whenever necessary against perennial weeds. Likewise, the application of organic fertilizers, manures and even green manures normally consists of at least shallow tillage operations. Thus, conservation tillage in organic agriculture poses some limitations in controlling weeds without herbicides, as well as in nutrient supply for reduced mineralization rate [3]. Still, it could provide positive effects on the environment.

At present, the effects of combined organic conservation systems have not been widely tested either from an agronomic or environmental point of view [10]. There is a lack of studies testing the effect of organic conservation agriculture on the emissions of carbon dioxide (CO_2), nitrous oxide (N_2O) and methane (CH_4) and the capability to mitigate GHG emissions compared to conventional systems.

In contrast, the effect of conservation tillage practices on GHG emissions from soil has been largely investigated, though with uncertain results. Indeed, some studies reported that conservative tillage increases N_2O and CH_4 emissions with respect to conventional tillage, while other studies reported lower GHG emissions in conservative than conventional tillage [11–13].

Concerning organic agriculture, soil N_2O emissions may be affected by the use of organic fertilizers. According to a recent meta-analysis, the use of organic fertilizers can significantly reduce N_2O emissions (23% reduction) in Mediterranean conditions with respect to the use of synthetic fertilizers [14].

Both conservation and organic agriculture are characterized by the use of cover crops, due to their well-known benefits for nutrient supply, organic carbon input and for the reduction of soil erosion and nitrate leaching risks [15], but their effect in terms of soil GHG mitigation was investigated only recently [16,17]. The inclusion of cover crops in crop rotations may mitigate soil GHG emissions thanks to an increase in carbon sequestration, a reduction of mineral fertilizers and a decrease in the N losses thanks to the uptake of nitrate by catch crops both in crop and intercrop periods. However, organic agriculture normally adopts the incorporation of soil of the cover crops as green manures that can provoke N_2O emissions peaks in the short term after tillage, especially in cases of N-rich cover crops (i.e., legumes) [18,19]. In contrast, conservation agriculture uses cover crops as living mulch, and so far, only one study investigated the effect of this practice on soil GHG emissions, reporting that living mulch can be a source of N_2O emissions [20].

The effect of organic conservation systems on the potential of soil to uptake CH_4 is not widely reported and data have been collected only in temperate areas. Six et al. [21] summarized these data and reported a greater CH_4 uptake under conservation agriculture than under conventional agriculture, that was attributed to the higher pore continuity and the presence of ecological niches for methanotrophic bacteria in conservation agriculture [22]. Indeed, some authors observed lower CH_4 uptake both in organic and in conservation agriculture than in conventional agriculture, since several conventional agricultural practices (e.g., mineral nitrogen fertilization, inversion tillage) have an adverse impact on the activity of CH_4 oxidizing bacteria [22,23]. Consequently, a system integrating organic and conservation agriculture entails the combination of many of the above reported agricultural practices that can affect soil GHG emissions in different way. The hypothesis of this research is that an organic conservation system (ORG+) may reduce soil emissions of N_2O, CH_4 and CO_2 compared to integrated farming (INT) and organic (ORG) systems, and to that aim, soil GHG fluxes were measured in a recently implemented two-year irrigated vegetable crop rotation in the Mediterranean.

2. Materials and Methods

2.1. Experimental Site Characterization

A two-year field experiment was conducted in the Pisa coastal plain (43° 40' N Lat; 10° 19' E Long; 1 m above mean sea level and 0% slope), at the "Enrico Avanzi" Centre for Agro-Environmental Research of the University of Pisa (Tuscany, Italy) on an irrigated vegetable crop rotation.

The climate there is typical of the north-Mediterranean area, characterized by a long-term average annual rainfall of 907 mm and a mean annual temperature of 15 °C (1986–2013).

The soil is a loamy sand originated from alluvial sediments and classified as a Typic Xerofluvent based on the USDA soil taxonomy [24]. At the beginning of the field experiment the soil was analysed at two depths (0–10 cm and 10–30 cm) to determine: Soil texture (international pipette method), pH (H_2O, 1:2.5), soil organic matter content (Walkley-Black method), total N content (Kjeldhal method), available P (Olsen method), exchangeable K ($BaCl_2$ method), conductivity (conductivity meter), C:N and bulk density (soil core method) (Table 1).

Table 1. Characterization of soil in the two fields (F1, F2) and at two depths (0–10 cm, 10–30 cm).

Parameter	Unit	Field 1		Field 2	
		0–10	10–30	0–10	10–30
Sand (2 mm–0.05 mm)	%	81.9	82.3	79.4	79.3
Silt (0.05 mm–0.002 mm)	%	13.6	12.6	14.4	13.9
Clay (< 0.002 mm)	%	4.5	5.1	6.2	6.8
pH	1:1 w/v	6.7	6.1	7.2	7.1
Organic Matter	%	2.2	1.9	2.6	2.2
Total N	$g\ kg^{-1}$	1.2	1.1	1.5	1.3
Available P	$mg\ kg^{-1}$	6.6	3.4	4.9	3.6
Exchangeable K	$mg\ kg^{-1}$	55.0	55.0	55.0	55.0
Conductivity	$\mu S/cm^{-3}$	153.3	82.7	185.6	88.4
C:N	-	10.8	10.4	9.9	10.0
Bulk density	$g\ cm^{-3}$	1.40		1.44	

The soil water table range from 70 cm during winter to 120 cm in summer.

2.2. Experimental Design and Management of the Cropping Systems

The field trial was set up in July 2014. The crops included in the rotation were: Savoy cabbage (*Brassica oleracea* var. sabauda L. cv. Famosa), spring lettuce (*Lactuca sativa* L. cv. Justine), fennel (*Foeniculum vulgare* Mill. Cv. Montebianco) and summer lettuce (*Lactuca sativa* L. cv. Ballerina) (Figure 1).

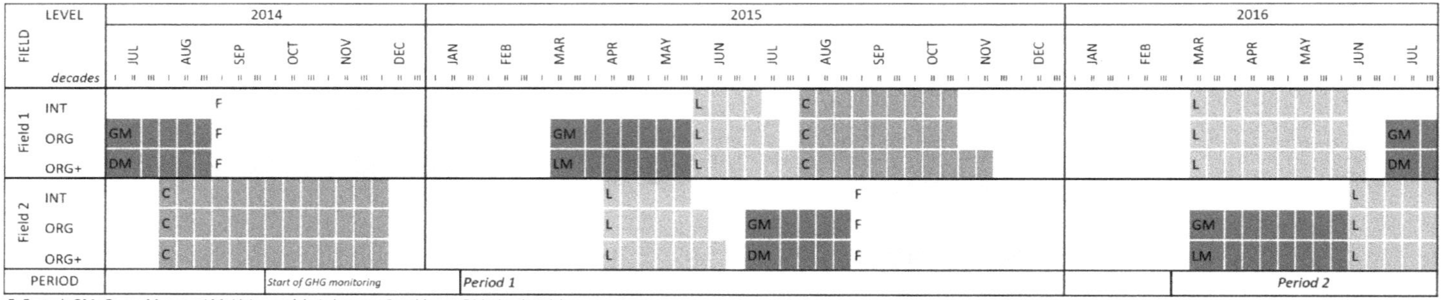

F: *Fennel; GM: Green Manure; LM: Living mulch; L: lettuce; C: cabbage; DM: dead mulch.*

Figure 1. Presence of the crops in rotation in the three cropping systems (integrated farming (INT), organic farming (ORG) and organic-conservation farming (ORG+)) in the two fields, with the calendar of soil greenhouse gas (GHG) monitoring. Soil GHG data from periods identified as Period 1 (P1) and Period 2 (P2) were used to calculate daily and cumulative emissions.

The two-year vegetable crop rotation was cultivated under three different management systems: integrated farming with conventional tillage practices, chemical pesticide uses and mineral fertilization (INT); organic farming with conventional tillage practices, organic fertilizers, green manure and physical (mechanical with roller crimper and thermal with flaming) weed control (ORG); organic farming combined with conservation practices including no-tillage, organic fertilizers and cultural weed control (ORG+).

The crop rotation was replicated in space and time. The spatial replicates were two adjacent fields: field 1 (F1), in which the rotation started with fennel, and field 2 (F2), in which the rotation started with cabbage. In each field, the three systems were completely randomized with three replicates constituted by an elementary plot of 3 m width × 21 m length.

The ORG system included a spring green manure mixture incorporated into the soil before transplanting of summer lettuce, composed of field peas (*Pisum sativum* L.) and faba beans (*Vicia faba* subsp. *minor* L.), and a summer green manure mixture—chopped and incorporated into the soil before fennel transplanting—composed of red cowpeas (*Vigna unguiculata* L. Walp), buckwheat (*Fagopyrum esculentum* L.), millet (*Panicum miliaceum* L.) and foxtail millet (*Setaria italica* L.). The ORG+ system included a red clover (*Trifolium pratense* L.) directly seeded and established as a living mulch for both summer lettuce and cabbage, and a summer dead mulch, terminated as dead mulch by roller crimper and flaming before the transplanting of fennel, composed of the same plants used in the spring green manure mixture of the ORG system.

Sprinkler irrigation was applied to all treatments during summer season (May–September). Irrigation was supplied daily in the ten days after transplant, and afterwards every 3 days until harvest. No irrigation was provided after significant rain events.

Potassium and phosphate fertilizers were provided just before transplanting (Table 2).

Total nitrogen fertilization of the three cropping systems for the two years was equal to 302.5 kg N ha^{-1} in INT from mineral fertilizers, 155.6 kg N ha^{-1} in ORG from organic fertilizers and 56 kg N ha^{-1} in ORG+ (organic fertilizers) (Tables 2 and 3).

The level of fertilization and application splits applied in the INT system were in compliance with the maximum amount of fertilizers stated by the integrated pest management (IPM) production disciplinary of Tuscan Regional Government . The fertilization strategy adopted in the ORG and ORG+ systems differed according to their respective references. The ORG system reproduced the standard organic management of field vegetables practiced by growers in the area. The level of fertilization was set as a trade-off between the target of achieving viable yields and keeping production costs under the threshold for profitability. The ORG+ was set as an agro-ecological system aimed at maximising the use of internal natural resources and the provision of agroecosystem services from cover crops (i.e., dead mulch and living mulch), whilst minimising negative impacts on the environment (e.g., by reducing soil tillage and external input application). That is why for ORG+ the level of fertilization was conceived as the minimum amount required by the crops, differentiated according to specific crop needs, to start growing after transplanting, while the remaining amount of nutrients has been assumed to be available from soil or cover crops. Detailed information about agricultural operations, fertilizations and weed management are reported in Tables 2 and 3.

Table 2. Agricultural practices carried out in field 1 for each crop in the three cropping systems; in field 2 the same agricultural practices were carried out, starting with savoy cabbage.

Data	Crop	Level	Main tillage	Sowing	Fertilization rate kg ha^{-1} of: N; P$_2$O$_5$; K$_2$O	Weed	Pest	Residue management
Jul–Jan	Fennel	INT ORG ORG+	Spading Spading No-till	Transplanting Transplanting No till transplanting	122; 138; 245 77; 94; 150 25; 58; 75	Chemical and mechanical weeding Mechanical weeding Flame weeding	Chemical	Removed Removed Removed
Feb–May	Spring green manure*	INT ORG ORG+	Rotary tiller, incorporation into the soil	Broadcast seeding				Chopped and incorporated into the soil with spade
Feb–May	Spring living mulch**	INT ORG ORG+		No till broadcast seeding				
Jun–Jul	Summer Lettuce	INT ORG ORG+	Spading Spading No-till	Transplanting Transplanting No till transplanting	46; 46; 110 0; 29; 75 0; 0; 0	Chemical and mechanical weeding Mechanical weeding Flame weeding	Chemical	Removed Removed Removed
Jul–Feb	Savoy cabbage	INT ORG ORG+	Spading Spading No-till	Transplanting Transplanting No till transplanting	108; 69; 173 59; 48; 96 28; 29; 50	Chemical and mechanical weeding Mechanical weeding Flame weeding	Chemical	Removed Removed Removed
Ma–May	Spring Lettuce	INT ORG ORG+	Spading Spading No-till	Transplanting Transplanting No till transplanting	27; 39; 75 20; 21; 64 0; 0; 0	Chemical and mechanical weeding Mechanical weeding Flame weeding	Chemical	Removed Removed Removed
Jun–Jul	Summer green manure***	INT ORG ORG+	Rotary tiller	Broadcast seeding				Chopped and incorporated into the soil with spade
Jun–Jul	Summer dead mulch***	INT ORG ORG+		No till broadcast seeding, devitalization with roller crimper and flaming				Rolled and band flamed

* field peas (*Pisum sativum* L.) and faba beans (*Vicia faba* subsp. minor L.); ** red clover (*Trifolium pratense* L.); *** red cowpeas (*Vigna unguiculata* L.), buckwheat (*Fagopyrum esculentum* L.), millet (*Panicum miliaceum* L.) and foxtail millet (*Setaria italica* L.).

Table 3. Type and splitting of fertilizers in the three cropping systems for each crop.

Crop	Level	Nitrogen fertilizer type and split
Fennel	INT	122 kg N ha^{-1} as ammonium nitrate 27% (A) - halved in two topdressing applications
	ORG	25.7 kg N ha^{-1} as a commercial fertilizer composed by a mixture of manures 5% N (B) - before transplanting 51.3 kg N ha^{-1} as blood meal fertilizer 14% (C) - halved in two topdressing applications
	ORG+	9.3 kg N ha^{-1} as B - before transplanting 18.7 kg N ha^{-1} as C - at transplanting
Summer Lettuce	INT	46 kg N ha^{-1} as A - halved in two topdressing applications
	ORG	
	ORG+	
Savoy cabbage	INT	108 kg N ha^{-1} as A - halved in two topdressing applications
	ORG	15 kg N ha^{-1} as B - before transplanting 44 kg N ha^{-1} as C - halved in two topdressing applications
	ORG+	7.5 kg N ha^{-1} as B - before transplanting 20 kg N ha^{-1} as C - halved in two topdressing applications
Spring Lettuce	INT	27 kg N ha^{-1} as A - halved in two topdressing applications
	ORG	19.6 kg N ha^{-1} as C - before transplanting
	ORG+	

2.3. Monitoring of Soil N_2O, CH_4 and CO_2 Flux

Fluxes of N_2O, CH_4 and CO_2 were measured from October 2014 to July 2016 by the flow-through non-steady state chamber technique [25], using a mobile instrument developed by West Systems Srl (Florence, Italy) within the LIFE+ "Improved flux Prototypes for N_2O emission reduction from Agriculture" (IPNOA) project (www.ipnoa.eu). The instrument is a light tracked vehicle that operates by remote control, equipped with a N_2O, carbon monoxide and water vapour detector that uses off-axis integrated cavity output spectroscopy (ICOS) and an ultraportable greenhouse gas analyser (UGGA) to measure CO_2, CH_4 and water vapour, both provided by Los Gatos Research (LGR) Inc. (Mountain View, CA, USA). Output gas concentrations are given with a scan rate of 1 s. Measured data were recorded using a smartphone connected via Bluetooth®. The technical details of the instrument and its validation were reported in Bosco et al. [26] and Laville et al. [27,28], respectively. Two PVC collars (15 cm height, 30 cm ∅) were inserted in each plot permanently at a soil depth of 5 cm and removed for short time only at the occurrence of tillage operations. The collars were mounted within plant rows and all the plants within the collars were removed by cutting the sprouts when necessary. To perform the flux measurement, a movable steel chamber (10 cm height, 30 cm ∅) was connected to the detector through a tube (20 m long, 4 mm ∅).

The chamber was equipped with an internal fan to guarantee the homogeneity of the gas concentration and a rubber seal to avoid air leaks. The deployment time of the chamber was 2–3 min.

The monitoring of soil GHG fluxes started on 10 October 2014 since the instrument was reserved for another field campaign. For the same reason the GHG monitoring campaign was interrupted from 18 December 2015 to 3 March 2016. Thus, for the calculation of cumulative GHG emissions and for the statistical analysis of the average daily fluxes the dataset was divided in two monitoring periods:

i. Period 1 (P1): going from 16 January 2015, the first day after the last harvest of the winter crops (fennel in F1), until 18 December 2015;

ii. Period 2 (P2): going from 3 March 2016 until the end of the monitoring campaign, 24 June 2016 in F1 and 14 July 2016 in F2.

2.4. Auxiliary Measurements

Daily air temperature, atmospheric pressure and rainfall were recorded from the closest weather station (less than 500 m).

Soil temperature and volumetric water content were measured close to each collar simultaneously with the measurement of GHG fluxes from soil, using a dielectric probe (Decagon Devices GS3) inserted into the soil at a depth of 5 cm and linked to the instrument via Bluetooth® connection. Soil water content values were used to calculate the soil water filled pore space (WFPS) according to Equations (1) and (2).

$$\text{Total porosity (\%)} = \frac{1 - \text{bulk density}}{2.65} \times 100 \tag{1}$$

$$\text{WFPS (\%)} = \frac{\text{volumetric water content}}{\text{total porosity}} \times 100 \tag{2}$$

In Equation (1), bulk density was measured using the soil core method and particle density was considered equal to 2.65 g cm^{-3}.

Soil samples were collected from the 0–20 cm soil layer for the determination of nitrate content (N–NO$_3$) in each plot. Three soil cores per plot were mixed to constitute one sample. The samples were stored at 4 °C before their analysis. Before the analysis, each soil sample was dried at 40 °C until constant weight and then it was sieved at 2 mm. A 10 g subsample of soil was extracted using deionised water in 1:2.5 ratio and then it was shacked for 120 min. N–NO$_3$ concentrations were determined using ionic chromatograph. Soil N–NO$_3$ content was calculated based on N sample concentration considering soil dry weight.

2.5. Data Elaboration and Statistical Analysis

Data elaboration and statistical analysis were performed with R software [29], considering $\alpha = 0.05$ as the passable level of significance.

N$_2$O, CH$_4$ and CO$_2$ measurements were checked for outliers among replicates in each sampling day, through the Grubbs test. After outlier removal, N$_2$O data were log transformed, as residuals deviated strongly from normal distribution. To enable this log-transformation, given the presence of negative values for daily N$_2$O fluxes, N$_2$O fluxes were translated before transformation as: $(N_2O\ flux + 0.1) - min\ (N_2O\ flux)$, where $min\ (N_2O\ flux)$ was the minimum value in the dataset.

One-way ANOVA was used to analyse the effect of the factor "system" in each sampling date and separately for the two fields on: GHG daily fluxes, soil temperature, soil WFPS and soil nitrate concentration along the overall monitoring campaign.

The effect of the systems on average daily fluxes was analysed in the two periods (P1 and P2) and for the two fields separately, through linear mixed effect models, one for each gas, using the R "lme4" package [30]. The two fields were analysed separately because each phase of the crop rotation did not occur simultaneously in the two fields, since the first crops in summer 2014 were fennel in F1 and cabbage in F2.

The system was considered as a fixed factor of the linear mixed effect models, with the replicate as a random effect. When the system had a significant effect on the studied variable, Tukey's HSD post hoc test ($\alpha = 0.05$) was used to reveal the differences between the levels of the factor system.

The relationships among soil temperatures WFPS, N$_2$O, CH$_4$ and CO$_2$ were analysed through the Spearman's correlation using the data collected across the overall field campaign and pooling the data of the two fields. Furthermore, the relationship between N$_2$O daily flux and soil nitrate concentration was evaluated through the Spearman's correlation, considering the monitoring days in which the soil samples were collected. The relationship between CO$_2$ emissions and soil temperature was evaluated to be exponential by plotting the data. Consequently, the analysis of covariance (ANCOVA) was used to compare the relationships between the logarithm of CO$_2$ flux and the soil temperature in the three levels of the factor "system".

Cumulative emissions of N$_2$O, CH$_4$ and CO$_2$, for both P1 and P2 were calculated by linear interpolation between two close sampling dates and the numerical integration of the function over time, assuming that fluxes changed linearly among sampling days. The effects of the system on the cumulative emissions were analysed through linear mixed effect models, which were built for each gas in the same way as for the daily fluxes.

The overall GHG budget (CO_2 equivalents) was calculated multiplying the cumulative value of each gas per period and field by the corresponding global warming potential (GWP) of AR5 [31]. The CO_2 equivalents (CO_2-eq) were calculated (i) separately for the non-CO_2 gases, as the sum of cumulative emissions of N_2O and CH_4, and (ii) as the net GHG emissions, also considering CO_2 emissions.

3. Results

3.1. Meteorological Conditions

During the GHG emissions monitoring periods, monthly average temperatures higher than 20 °C were recorded in summer 2015 (average of June, July and August 24 °C) and summer 2016 (average of June and July 22 °C) (Figure 2).

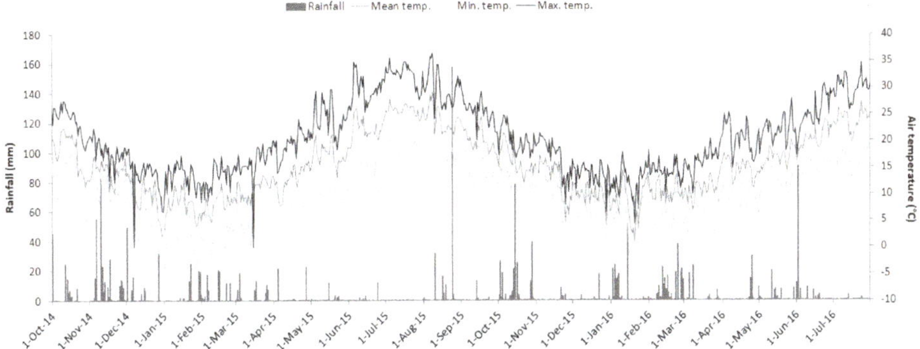

Figure 2. Daily rainfall (mm), daily maximum, average and minimum air temperature (°C) from October 2014 to July 2016.

The monthly average temperature was lower than 10 °C in January–February 2015 and January 2016 (8 °C). The rainiest month was November 2014 (290 mm), while the driest month was July 2015 (3 mm). Along the whole monitoring period, the rainiest periods were August 2015 (232 mm), October 2015 (254 mm), the period between January and February 2016 (372 mm) and in June 2016 (138 mm).

3.2. Soil Water Content, Temperature and Nitrate Dynamic

Water filled pore space (WFPS) values did not differ significantly among INT, ORG and ORG+ systems, in either in F1 or F2, with exceptions of (i) 20 May 2015 in F1, where ORG+ and INT had higher WFPS than ORG, and (ii) the period between May and June 2016 in F2, where ORG+ showed significantly higher WFPS values (Figures 3a and 4a).

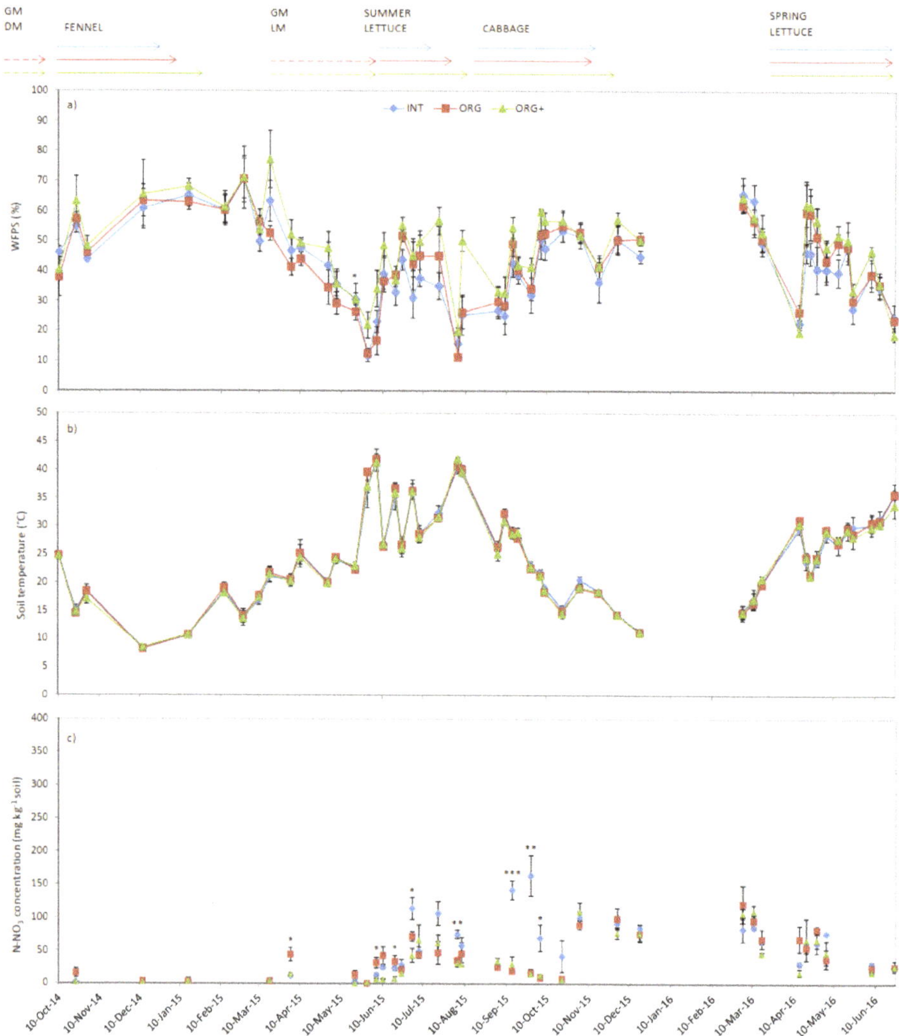

Figure 3. Data recorded in F1: (**a**) Soil water filled pore space (WFPS); (**b**) soil temperature; (**c**) soil nitrate (N–NO$_3$) concentration for each treatment. Simple arrows indicate fertilization events, and dashed arrows the primary tillage of each crop. On field 1 (F1) the temporal crop sequence was: Fennel, summer lettuce, cabbage, then spring lettuce. Significance was as follows: *n.s.* is not significant; * is significant at the $p \leq 0.05$ level; ** is significant at $p \leq 0.01$ level; *** is significant at $p \leq 0.001$ level.

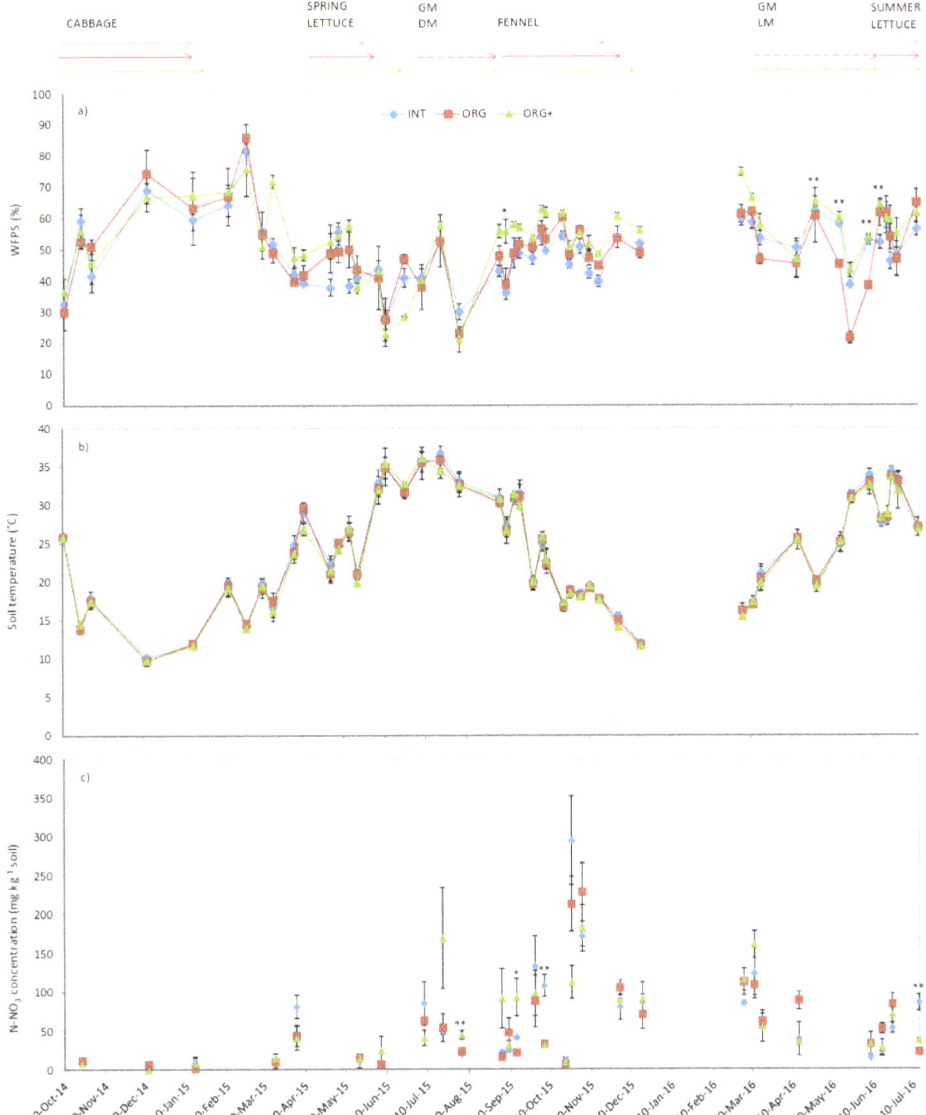

Figure 4. Data recorded in F2: (**a**) Soil WFPS; (**b**) soil temperature; (**c**) soil nitrate (N–NO$_3$) concentration for each treatment. Simple arrows indicate fertilization events; dashed arrows the primary tillage of each crop. On field 1 (F1) the temporal crop sequence was: Cabbage, spring lettuce, fennel, then summer lettuce. Significance was as follows: *n.s.* is not significant; * is significant at the $p \leq 0.05$ level; ** is significant at $p \leq 0.01$ level.

The highest WFPS values were registered in both fields in winter, with maximum values in February 2015 (71% in F1 and 81% in F2) and minimum values in May 2015 (12% in F1 and 23% in F2). Indeed, average WFPS values were high in summer period due to irrigation (36% in F1 and to 46% in F2).

Soil temperature was not different among treatments in both fields. The lowest soil temperature (9 °C) was recorded in December 2014 and the highest soil temperature (39 °C) in June and August 2015 (Figures 3b and 4b).

Soil nitrate concentration showed values ranging from 0 to 163 mg kg^{-1} in F1 and up to 295 mg kg^{-1} in F2. Nitrate concentration was higher than 60 mg N–NO$_3$ kg^{-1} in 17 sampling dates out of 33 in F1 and in 17 sampling dates out of 28 in F2. In F1, nitrate concentrations were significantly higher in INT than the other treatments in five dates from July 2015 to October 2015 (average 112 mg N–NO$_3$kg^{-1}); and in ORG in three dates in April 2015 and in June 2015, with summer lettuce (average 36.2 mg N–NO$_3$kg^{-1}). In F2, nitrate concentration was higher in cabbage INT on one date in October 2015 (107.8 mg N–NO$_3$N kg^{-1}) and on one date in July 2016 (average 85.4 mg N–NO$_3$ kg^{-1}). It was higher in ORG+ during August and September 2015; in this case after organic nitrogen fertilization for cabbage (93.1 mg N–NO$_3$ kg^{-1}) (Figures 3c and 4c).

3.3. Daily Flux of N$_2$O, CH$_4$ and CO$_2$

Pattern of N$_2$O, CH$_4$ and CO$_2$ fluxes throughout the study period are show in Figure 5a, b, c for F1 and Figure 6a, b, c for F2, while the ANOVA results are reported in Table 4.

3.3.1. Trend of Daily N$_2$O Flux in the Three Cropping Systems

Measured N$_2$O daily flux ranged from −0.4 to 53.3 mg N$_2$O m^{-2} day^{-1} in F1 and from −1.7 to 20.2 mg N$_2$O m^{-2} day^{-1} in F2 (Figures 5a and 6a). Notably, high N$_2$O fluxes were observed in F1 in June 2015 in ORG system after green manure incorporation into the soil (20.2 mg N$_2$O m^{-2} day^{-1}), in August 2015 (53.3 mg N$_2$O m^{-2} day^{-1}) in the ORG+ system just after organic fertilization on cabbage, and in April 2016 in the ORG system (37.3 mg N$_2$O m^{-2} day^{-1}) after tillage and organic nitrogen fertilization for spring lettuce.

In F2, N$_2$O peaks were halved compared to F1 and the highest were registered after the organic nitrogen fertilization of fennel in September 2015 on ORG+ (16.4 mg N$_2$O m^{-2} day^{-1}), in October 2015 on ORG (on average 15.5 mg N$_2$O m^{-2} day^{-1}) and after green manure incorporation into soil in June 2016 in ORG (8.3 mg N$_2$O m^{-2} day^{-1}).

In P1 (Jan 2015-Dec 2015), average daily N$_2$O flux (Table 4) in F1 was significantly lower in ORG+ (2.21 ± 1.18 mg N$_2$O m^{-2} day^{-1}), while no differences were observed between INT and ORG (on average 2.85 ± 0.32 mg N$_2$O m^{-2} day^{-1}). In F2, no differences were detected among the three cropping systems (on average 2.36 ± 0.29 mg N$_2$O m^{-2} day^{-1}).

During P2 (Jan 2016–Jul 2016), in F1 the effect of the cropping systems on the average daily N$_2$O flux was the same as that in P1, with INT equal to ORG, and the highest values were recorded (on average 3.89 ± 1.15 mg N$_2$O m^{-2} day^{-1}) and ORG+ with the lowest value (0.47 ± 0.12 mg N$_2$O m^{-2} day^{-1}). In F2 N$_2$O daily flux was significantly higher in ORG (2.63 ± 0.59 mg N$_2$O m^{-2} day^{-1}) than in ORG+ (1.39 ± 0.52 mg N$_2$O m^{-2} day^{-1}).

3.3.2. Trend of Daily CH$_4$ Flux in the Three Cropping Systems

Measured CH$_4$ daily flux ranged from –0.7 to 0.45 mg CH$_4$ m^{-2} day^{-1} in F1 and from −0.47 to 0.43 mg CH$_4$ m^{-2} day^{-1} in F2 (Figures 5b and 6b). In F1, CH$_4$ fluxes were positive (<0.2 mg CH$_4$ m^{-2} day^{-1}) in 12, 9 and 11 sampling days out of 50 in INT, ORG and ORG+, respectively. In F2, CH$_4$ fluxes were positive (<0.5 mg CH$_4$ m^{-2} day^{-1}) in seven, two and six sampling days out of 48 in INT, ORG and ORG+, respectively. In particular, in F2, CH$_4$ fluxes were significantly lower in ORG than in INT and ORG+ in two sampling dates in March 2016 and in May 2016; during which CH$_4$ fluxes in ORG were equal to −0.35 ± 0.04 mg CH$_4$ m^{-2} day^{-1} and −0.34 ± 0.07 mg CH$_4$ m^{-2} day^{-1}, respectively. In F1 the average daily CH$_4$ flux (Table 4) was slightly negative, with no significant differences ($p > 0.05$) among the cropping systems in both periods (on average P1: −0.10 ±0.22 mg CH$_4$ m^{-2} day^{-1}; P2: −0.08 ± 0.02 mg CH$_4$ m^{-2} day^{-1}). In F2, significantly lower flux was recorded in P1 in ORG (−0.23 ± 0.02 mg CH$_4$ m^{-2} day^{-1}) compared to INT and ORG+ (−0.18 ± 0.01 mg CH$_4$ m^{-2} day^{-1}), while in P2 ORG and

ORG+ showed similar values equal to -0.10 ± 0.02 mg CH_4 m^{-2} day^{-1}, significantly lower than INT (-0.001 ± 0.05 mg CH_4 m^{-2} day^{-1}).

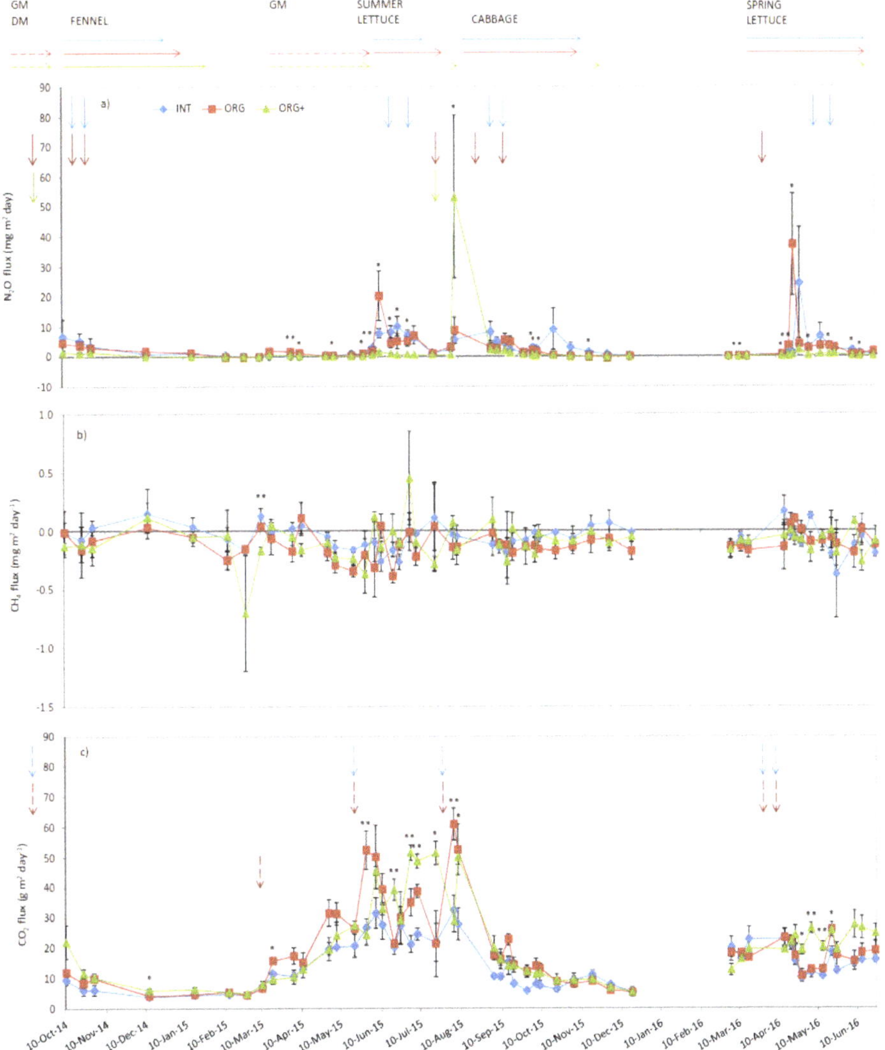

Figure 5. Daily average fluxes recorded in F1 of: (**a**) N_2O; (**b**) CH_4; and (**c**) CO_2 for each treatment. Simple arrows indicate fertilization events; dashed arrows the primary tillage of each crop. On field 1 (F1) the temporal crop sequence was: Fennel, summer lettuce, cabbage, then spring lettuce. Significance was as follows: *n.s.* is not significant; * is significant at the $p \leq 0.05$ level; ** is significant at $p \leq 0.01$ level.

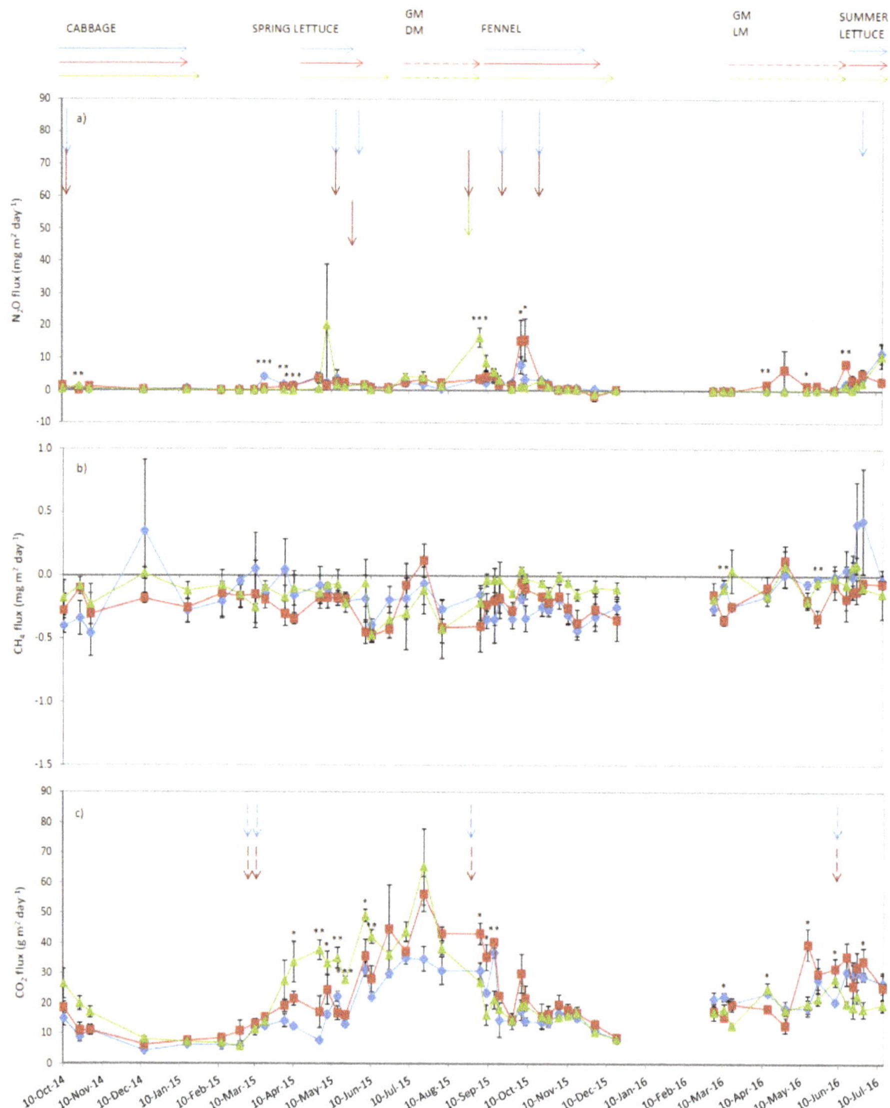

Figure 6. Daily average fluxes recorded in F2 of: (**a**) N_2O; (**b**) CH_4; and (**c**) CO_2 for each treatment. Simple arrows indicate fertilization events; dashed arrows the primary tillage of each crop. On field 1 (F1) the temporal crop sequence was: Cabbage, spring lettuce, fennel, then summer lettuce. Significance was as follows: *n.s.* is not significant; * is significant at the $p \leq 0.05$ level; ** is significant at $p \leq 0.01$ level; *** is significant at $p \leq 0.001$ level.

Table 4. Effects of the three cropping systems on average daily flux of CO_2, CH_4 and N_2O, during the two monitoring periods (P1: January 2015–December 2015; P2: January 2016–July 2016) in Field 1 and Field 2. System levels are INT: Integrated; ORG: Organic; ORG+: Conservation organic. Different letters represent significant differences between the cropping systems resulting from the post-hoc test. Values are mean $\pm$ SE; $n = 18$.

Period		F1			F2		
		N_2O (mg m^{-2} day^{-1})	CH_4 (mg m^{-2} day^{-1})	CO_2 (g m^{-2} day^{-1})	N_2O (mg m^{-2} day^{-1})	CH_4 (mg m^{-2} day^{-1})	CO_2 (g m^{-2} day^{-1})
P1	System	$p < 0.0001$	n.s.	$p < 0.0001$	n.s.	$p < 0.001$	$p < 0.0001$
	INT	3.05 ± 0.40 a	-0.06 ± 0.02	15.11 ± 0.98 b	2.07 ± 0.23	-0.21 ± 0.02 a	18.08 ± 1.00 b
	ORG	2.66 ± 0.49 a	-0.08 ± 0.06	22.09 ± 1.72 a	2.45 ± 0.47	-0.23 ± 0.02 b	23.63 ± 1.42 a
	ORG+	2.21 ± 1.18 b	-0.10 ± 0.03	20.93 ± 1.58 a	2.56 ± 0.70	-0.14 ± 0.02 a	24.12 ± 1.54 a
P2	System	$p < 0.0001$	n.s.	$p < 0.0001$	$p < 0.05$	$p < 0.05$	$p < 0.0001$
	INT	3.37 ± 1.47 a	-0.09 ± 0.03	15.59 ± 0.97 b	1.62 ± 0.55 ab	0.00 ± 0.05 a	24.66 ± 0.96 a
	ORG	4.40 ± 1.77 a	-0.07 ± 0.03	16.59 ± 1.00 ab	2.63 ± 0.59 a	-0.14 ± 0.03 b	26.11 ± 1.66 a
	ORG+	0.47 ± 0.12 b	-0.08 ± 0.02	19.81 ± 1.05 a	1.39 ± 0.52 b	-0.05 ± 0.03 b	19.97 ± 0.79 b

3.3.3. Trend of Daily CO_2 Flux in the Three Cropping Systems

Measured CO_2 daily flux ranged from 3.9 to 60.9 g CO_2 m^{-2} day^{-1} in F1 and from 4.4 to 65.2 g CO_2 m^{-2} day^{-1} in F2 (Figures 5c and 6c). Daily pattern of CO_2 flux varied according to that of soil temperatures. Indeed, higher values of CO_2 flux were recorded from May 2015 to September 2015. Higher CO_2 flux was observed in ORG+ than in the other systems during summer 2015 in both fields. In F1, significantly higher emissions were observed in ORG+ with respect to the other treatments in eight dates out of 50, while CO_2 flux was higher in ORG than in the other systems in five dates out of 50. In F2, CO_2 flux was significantly higher in nine dates out of 48 in ORG+ and in six dates out of 48 in ORG. Higher CO_2 fluxes in ORG systems were observed in March 2015 after tillage for green manure sowing, and in summer 2015, some days after main tillage operations for summer lettuce, and cabbage in F1, and for fennel in F2. Otherwise, CO_2 flux was significantly higher in INT than in the other treatments in only two dates in March 2016 in F2.

In both fields, daily average CO_2 flux (Table 4) in P1 was higher in ORG and ORG+ (on average, F1: 21.51 ± 1.16 g CO_2 m^{-2} day^{-1}; F2: 23.88 ± 1.05 g CO_2 m^{-2} day^{-1}) than in INT (F1: 15.11 ± 0.98 g CO_2 m^{-2} day^{-1}, F2: 18.08 ± 1.00 g CO_2 m^{-2} day^{-1}) In P2 higher CO_2 flux was higher in ORG+ in F1 compared to INT, while the opposite was recorded in F2, where INT and ORG (on average 25.40 ± 1.31 g CO_2 m^{-2} day^{-1}) recorded higher values than ORG+ (19.97 ± 0.79 g CO_2 m^{-2} day^{-1}).

3.4. Relationship among the Soil Variables and GHG Fluxes

The correlation between daily N_2O flux and soil N–NO$_3$ concentration-computed using a subset of the dataset, including only the monitoring days in which the soil samples were collected-turned out to be non-significant for all the three cropping systems (data not shown).

Considering the whole dataset, soil temperature and WFPS correlated negatively in all the three cropping systems, with a correlation coefficient (r_s) between −0.55 (ORG+) and −0.65 (INT) (Figure 7).

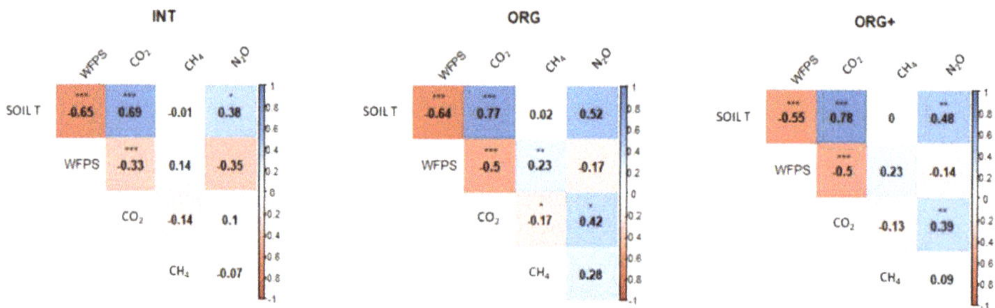

Figure 7. Correlation plot among the soil variables and GHG fluxes. Numbers indicate the correlation coefficients, while the intensity of the colour of the boxes represents the level of correlation according to the scale reported close to each plot. Significance was as follows: * is significant at the p ≤ 0.05 level; ** is significant at p ≤ 0.01 level; *** is significant at p ≤ 0.001 level.

Flux of N_2O correlated positively with soil temperature in INT (r_s: 0.38) and ORG+ (r_s: 0.48); and with CO_2 flux in ORG (r_s: 0.42) and ORG+ (r_s: 0.39).

Flux of CO_2 correlated positively with soil temperature with r_s equal to 0.69 in INT, 0.77 in ORG and 0.78 in ORG+; and negatively with WFPS with r_s between −0.33 (INT) and −0.5 (ORG and ORG+). Flux of CH_4 correlated positively with WFPS only in ORG (r_s: 0.23) and negatively with CO_2 flux in the same treatment (r_s: −0.17).

Fluxes of N_2O higher than 20 mg m^{-2} day^{-1} were recorded with WFPS values between 38% and 70% (Figure S1a). When WFPS was lower than 38% N_2O fluxes ranged between −0.04 mg m^{-2} day^{-1} and 17.44 mg m^{-2} day^{-1}, while when WFPS values were higher than 70%, N_2O fluxes ranged between −0.07 mg m^{-2} day^{-1} and 3.35 mg m^{-2} day^{-1}.

The ANOVA describing the relationship between the logarithm of CO_2 flux and the soil temperature highlighted that the slope of the linear regression was not different according to the treatments (0.059), while the intercept of the regression was significantly lower in INT (1.249) than in ORG (1.471) and ORG+ (1.487) (Figure 8).

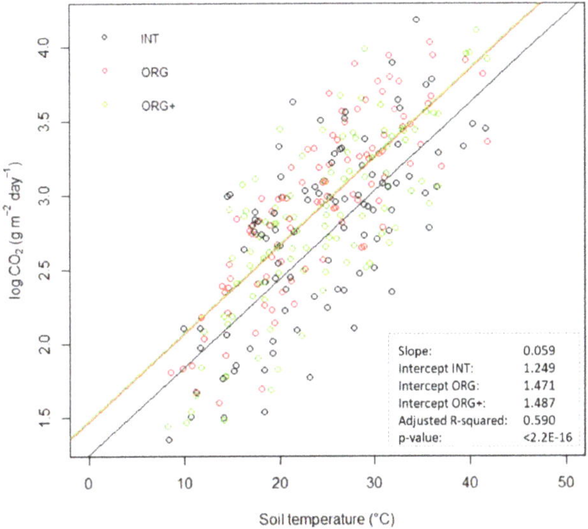

Figure 8. Relationship between the logarithm of CO_2 flux and the soil temperature.

3.5. Cumulative Soil Emissions during the Two Periods

In P1 cumulative N_2O emissions showed no significant differences among the cropping systems both in F1 (average 5.5 ± 1.1 kg N–N_2O ha^{-1}) and in F2 (average 4.4 ± 0.5 kg N–N_2O ha^{-1}) (Figure 9a).

In P2 cumulative N_2O emissions were significantly affected by the cropping system in both fields ($p < 0.05$). Indeed, in F1, cumulative N_2O emissions were higher in INT and in ORG (average 2.0 ± 0.5 kg N–N_2O ha^{-1}) than in ORG+ (0.3 ± 0.1 kg N–N_2O ha^{-1}). In F2 N_2O emissions were significantly higher in ORG (2.2 ± 0.7 kg N–N_2O ha^{-1}) than in ORG+ 1.0 ± 0.2 kg N–N_2O ha^{-1}), and INT was not significantly different from both ORG and ORG+ (1.1 ± 0.1 kg N–N_2O ha^{-1}).

There was an overall sink effect for CH_4 cumulative emissions in all systems, in both periods and fields, with no significant differences among cropping systems (average in F1: -162 ± 38 g C–CH_4 ha^{-1}; average in F2: -356 ± 60 g C–CH_4 ha^{-1}) (Figure 9b).

Cumulative CO_2 emissions in P1 were significantly affected by cropping system in both fields (F1: $p < 0.01$; F2: $p < 0.05$) (Figure 9c). Lower values were recorded in both fields in INT (F1: 13.0 ± 1.0 t C–CO_2 ha^{-1}; F2: 16.7 ± 0.8 t C–CO_2 ha^{-1}) than in ORG and ORG+ (average in F1: 18.3 ± 0.7 t C–CO_2 ha^{-1}; average in F2: 22.6 ± 0.6 t C–CO_2 ha^{-1}). In P2 differences were significant only in F1 ($p < 0.01$), where cumulative CO_2 emissions were higher in ORG+ (8.5 ± 0.4 t C–CO_2 ha^{-1}) than in ORG and INT (5.4 ± 0.3 t C–CO_2 ha^{-1}).

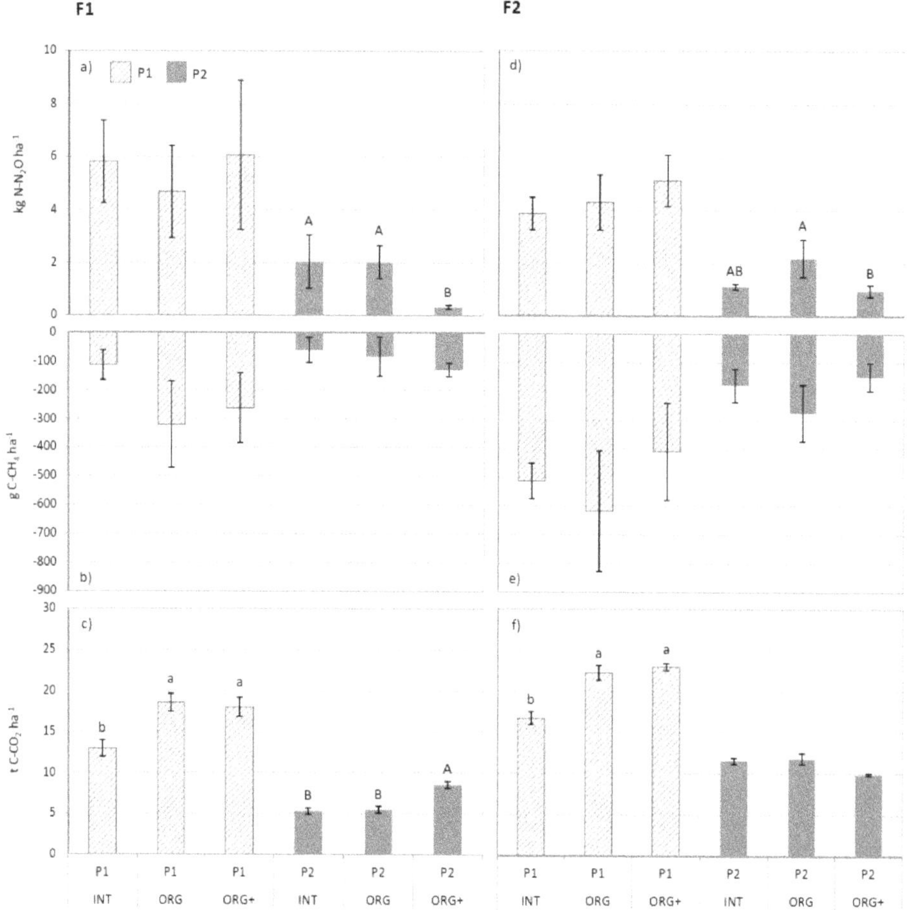

Figure 9. Cumulative emissions of N$_2$O (kg N–N$_2$O ha^{-1}), CH$_4$ (g C–CH$_4$ ha^{-1}) and CO$_2$ (t C–CO$_2$ ha^{-1}) for P1 and P2 in field 1 (**a, b, c**) and in field 2 (**d, e, f**), respectively. Different lowercase letters in P1, and uppercase letters in P2 indicate significant differences between the cropping systems resulting from the post-hoc test.

The estimated net GHG emissions (CO$_2$-eq) were significantly affected by the cropping systems with exception of P2 in F2 (Table 5).

Table 5. Estimated cumulative CO_2 emissions (t CO_2 ha^{-1}): CO_2 equivalents of non-CO_2 GHG as the sum of cumulative N_2O and CH_4 emissions (t CO_2-eq ha^{-1}), and net GHG emissions (t CO_2-eq ha^{-1}) during the two monitoring periods (P1: January 2015-December 2015; P2: January 2016– July 2016) in Field 1 and Field 2. System levels are INT: Integrated; ORG: Organic; ORG+, conservation organic. Different letters represent significant differences between the cropping systems resulting from the post-hoc test.

Period		CO_2 emissions (t CO_2 ha^{-1})		CO_2 equivalents of non-CO_2 GHG (t CO_2-eq ha^{-1})		Total CO_2-equivalents (t CO_2-eq ha^{-1})	
		F1	F2	F1	F2	F1	F2
P1	System	$p < 0.001$	$p < 0.01$	*n.s.*	*n.s.*	$p < 0.01$	$p < 0.01$
	INT	47.5 ± 3.6 b	61.3 ± 2.8 b	2.7 ± 0.73	1.8 ± 0.29	50.2 ± 4.1 b	63.1 ± 2.8 b
	ORG	68.1 ± 4.0 a	81.6 ± 3.3 a	2.2 ± 0.82	2.0 ± 0.50	70.2 ± 4.8 a	83.6 ± 3.5 a
	ORG+	66.0 ± 4.4 a	84.4 ± 3.1 a	2.8 ± 1.32	2.4 ± 0.46	68.8 ± 5.1 a	86.8 ± 3.2 a
P2	System	$p < 0.001$	*n.s.*	*n.s.*	*n.s.*	$p < 0.01$	*n.s.*
	INT	19.5 ± 1.4 b	42.4 ± 1.4	1.0 ± 0.47	0.5 ± 0.05	20.5 ± 1.8 b	42.9 ± 1.3
	ORG	20.2 ± 1.5 b	43.4 ± 2.4	0.9 ± 0.29	1.0 ± 0.33	21.2 ± 1.8 b	44.4 ± 2.7
	ORG+	31.3 ± 1.5 a	36.4 ± 0.5	0.1 ± 0.03	0.4 ± 0.11	31.5 ± 1.5 a	36.9 ± 0.5

In P1 in both F1 and F2 the net CO_2-eq were significantly higher in ORG (+40%, +33%), and ORG+ (+37%) than in INT. The CO_2-eq of non-CO_2 GHG were not different among INT, ORG and ORG+ in both fields and periods.

4. Discussion

This study evaluated the effect on GHG emissions from soil under three different agricultural management systems, an integrated (INT), an organic (ORG) and an organic conservation (ORG+) system, on an irrigated vegetable crop rotation for two years, and the relationship of GHG fluxes with soil variables.

Daily fluxes of N_2O correlated positively with soil temperature and CO_2 fluxes, probably caused by the high microbial activity associated to the organic matter mineralization in the warm season. Indeed, higher peaks in N_2O emissions occurred mainly between the end of March and the beginning of October, namely the period with higher soil temperatures (>20 °C). Other studies reported that soil temperature may be a driver for N_2O production when substrates are abundant, and the soil water content is optimal for microbial processes [32,33]. However, in our experiment the period with higher soil temperatures corresponded to that during which all N fertilization occurred, thus, it is difficult to consider separately the effect of the two drivers on N_2O emissions.

The agricultural management system influenced the average daily N_2O flux within F1 in P1 and P2, and within F2 in P2; in those cases, we found lower values in ORG+ than in the other systems, likely due to the significantly lower N fertilizer rate supplied to ORG+. Indeed, ORG+ had significantly lower cumulative N_2O emissions than INT and ORG in both fields in period 2, in which no fertilizers were supplied to spring and summer lettuce in ORG+.

We did not find a significant correlation between nitrate concentration in soil and N_2O emissions, even if nitrates were higher after mineral fertilization events in INT than organic fertilization in ORG and ORG+ in few sampling days in summer 2015, since the low number of the soil samples (29 in F1 and 27 in F2) may have negatively affected the robustness of the model.

The effect of nitrogen fertilization events, implying either mineral or organic N forms, on stimulating both short-term N_2O flux and cumulative N_2O emissions, was already reported by many authors [34,35]. In our study, high peaks of N_2O (>10 mg N_2O m^{-2} day^{-1}) were recorded a few days after fertilization events (4–10 days), in accordance to what was reported by Volpi et al. [36] in a similar soil and in the same environment.

In our study, peaks on daily N_2O flux were generally higher (>15 mg N_2O m^{-2} day^{-1}) after organic N fertilization events (ORG and ORG+), than after mineral N fertilization (INT). The occurrence of peaks in soil N_2O emissions after the application of organic fertilizers have been explained by other studies [37,38], as an effect of the increased availability of N and C for the soil microbial community. Thus, the increased microbial activity leads to high O_2 consumption that may create anaerobic conditions suited for the denitrification process from which N_2O is originated.

Differently, other studies reported lower N_2O emissions with organic fertilizers than mineral fertilizers, especially with solid manure, due to a slower release of N respect to mineral fertilizers or liquid slurry [14]. However, the effect of fertilizers on soil GHG emissions strictly depends on climate and soil specific conditions as well as on the type of the organic fertilizer itself. In fact, Pelster et al. [39], comparing four different N sources (one mineral fertilizer and three different manures), observed that N_2O emissions responded similarly to organic and mineral N sources in high-C soils, whereas in low-C soils N_2O emissions may be specifically stimulated by the use of C-rich manures. Moreover, the application technique of organic fertilizers may influence the soil N_2O emissions. Indeed, the incorporation of organic fertilizers is expected to increase N_2O emissions when soil moisture status is suitable for N_2O production, while ammonia volatilization may decrease, since more N entered the soil [40]. However, in our experiment we highlighted a tendency for lower N_2O emissions in ORG+ where the fertilizers were broadcasted more on soil surface than in ORG, where they were incorporated in soil, though that result was most probably due to the low N rate applied in ORG+.

Moreover, peaks of N_2O emissions, in a range between 5 and 20 mg N_2O m^{-2} day^{-1}, occurred from 10 to 15 days after the soil incorporation of the green manures in ORG. Heller et al. [41] in Mediterranean conditions, recorded the highest N_2O flux maximum two weeks after the tillage operations practiced for maize residues incorporation. Other authors reported that the incorporation of crop biomass into the soil produced N_2O and CO_2 peaks due to the increased availability of substrates for mineralization and microbial activity, when soil moisture was not limiting [42,43]. In particular, it was reported that N_2O emissions are generally increased when crop biomass with a low C:N ratio is incorporated in the soil [17,18]. However, in our study, peaks in N_2O emissions were similar (15–20 mg N_2O m^{-2} day^{-1}) after the incorporation of both spring green manure, composed by only legumes (C:N = 12, Tables S1 and S2) and summer green manure, composed by one legume, two cereals and one pseudo-cereal (C:N = 33, Tables S1 and S2). Indeed, peaks in N_2O emissions might have been due to an improvement of C availability in soil after plant material incorporation that stimulated denitrification [44].

Daily fluxes of CH_4 were negative in about 80% and 90% of the sampling days in F1 and F2, respectively. CH_4 uptake by soil was similar in all the cropping systems, with higher uptakes recorded only in the ORG system in F2 (average −0.19 mg CH_4 m^{-2} day^{-1}). Values of CH_4 uptake recorded in our experiment were in the range reported by literature for non-flooded agricultural soils (from 0 to 1.03 mg CH_4 m^{-2} day^{-1}) [22]. However, CH_4 uptake was lower than that reported by Flessa et al. [45] on a potato field in a temperate climate (average −0.35 mg CH_4 m^{-2} day^{-1}). Cumulative CH_4 emissions were not different among the cropping systems, in both periods and fields. In that regard, our results comply with other studies that reported no effect by conservation tillage on CH_4 emissions [46]. Differently, other studies comparing organic and non-organic management revealed a slightly, but significantly higher, net CH_4 uptake in organic cropping systems [47]. Moreover, the higher mineral fertilizer rate distributed in INT and the higher tillage intensity of INT and ORG seemed to have not inhibited the soil CH_4 oxidation capacity; namely the methanotrophic activity of microorganisms in soil, compared to the ORG+ system. However, the recent implementation of the three cropping systems could not yet have affected the soil stability, as well as the gas diffusion and the methanotrophic activity in soil that may influence CH_4 uptake [22]. Indeed, the number of years since the initiation of conservation tillage is a key issue for evaluating and understanding the effects generated by this management strategy [48].

Moreover, our study showed no differences in CH_4 emissions during periods of bare fallow (INT) and periods with cover crops (ORG and ORG+), similarly to what was reported by Sanz-Cobena et al. [49] and Guardia et al. [50] in a maize/cover crop rotation. However, studies are scarce on this topic, thus further research is needed to investigate the effect of cover crops on CH_4 emissions [16].

Concerning soil conditions, we did not find any strong correlations among soil temperature, WFPS and CH_4 emissions, with only a weak positive correlation between soil CH_4 emissions and WFPS in ORG. In our experiment WFPS did not show prevalently very low or high values, and soil water content was not as a strong driver for CH_4 emissions as reported in other studies, where it lowered the activity of methanotrophic bacteria in very dry or very wet soil conditions [51].

Measurements of daily CO_2 flux in our experiment ranged from 3.9 to 65.2 g CO_2 m^{-2} day^{-1}, with values often higher than in other studies conducted in a Mediterranean environment on fertilized crops, including organic cultivation or cover crops (1.5–25.7 g CO_2 m^{-2} day^{-1}) [49,52]. In all treatments, the intensity of CO_2 daily fluxes followed the variations of soil temperature, with values generally higher (up to 60 g CO_2 m^{-2} day^{-1}) during the warm season, between April and September, than in the rest of the year (<25 g CO_2 m^{-2} day^{-1}). Our results confirm the positive relationship between soil temperature and CO_2 flux, usually non-linear, reported by other authors [53–55]. In our experiment, irrigation may have contributed to the high values of CO_2 daily flux measured during the warm season, compared to those of other studies conducted in drought stressed Mediterranean environments. Indeed, Almagro et al. [56] reported that soil respiration varied following changes in soil moisture in late spring and summer, in a dry meso-Mediterranean climate, and that soil respiration was strongly limited by soil water content (SWC) < 10%. In our study, irrigation allowed us to maintain soil water content above 9% (20% WFPS), with the exception of three dates. In such a condition, soil water was never limiting for biological processes deputed to the production of CO_2, including root respiration. Our results highlighted a negative correlation between WFPS and CO_2 daily flux, only due to the stronger positive correlation of CO_2 daily flux and soil temperature and to the inverse pattern of WFPS and soil temperature values both in winter and in summer periods.

Furthermore, our results showed a different effect of the cropping systems on daily flux of CO_2, as the intercept of the linear regression describing the relationship between CO_2 flux and soil temperature was higher in ORG and ORG+ than in INT. Thus, besides the variation mediated by soil temperature and water content, the level of organic substrates supplied to the soil in ORG and ORG+ have determined higher soil respiration rates.

Moreover, the incorporation of soil of green manure (ORG) might have been a significant driver for short-term CO_2 fluxes, due to the proneness of green manure to mineralization [52,57]. Indeed, CO_2 daily flux was higher in ORG than in the other treatments a few days after the green manure incorporation was carried out, before summer lettuce cultivation in F1, and before fennel cultivation in F2.

Short-term peaks in CO_2 daily flux were also recorded in F1 after main tillage for sowing green manure and cabbage transplanting, and in F2 after tillage for fennel transplanting. Peaks in CO_2 emissions after main tillage were previously reported by many authors, mainly due to an increased mineralization of soil organic matter, as well as a transitory effect, due to the removal of physical constraints on CO_2 diffusion [46,58]. Cumulative CO_2 emissions ranged between 2.0 and 8.2 t C–CO_2 ha^{-1} and when there was a significant difference among the cropping systems, we highlighted a tendency in higher emissions in ORG and/or ORG+ than in INT. In ORG, the green manure incorporation and the organic fertilizer application could have increased the soil heterotrophic respiration [41] as discussed above, while in ORG+, living mulch may have increased the autotrophic component of respiration [59,60]. These results are in line with Chirinda et al. [61] that reported an increase of CO_2 emissions due both to manure application and to catch crops' cultivation in a sandy loam soil.

The net GHG emissions budget showed a tendency of being higher in ORG+ (in both fields and periods) and ORG (in P1 in F1 and F2) with respect to INT because of the effect of the cropping system on CO_2 emissions, since the CO_2-eq of non-CO_2 GHG were not different among INT, ORG and ORG+.

Thus, the integration of organic and conservation agriculture showed a tendency of higher CO_2 emissions and lower N_2O emissions than the other cropping systems, with no clear potential for soil GHG mitigation, at least in the first two years of organic conservation management. Indeed, a long-term field trial could help to clarify whether the result of this study on the effect of ORG+ on soil CO_2 and N_2O emissions was only transitory, especially considering the importance of the duration of no-till [62]. It is well known, indeed, that the introduction of no-till practices may require a long time to produce beneficial effects on soil's physical and biological aspects, which may buffer the GHG emission potential of the soil.

In the transition phase, a possible solution to improve the distribution of fertilisers in the soil profile and to sustain crop yield could come from a different fertilization strategy. The within-furrow application of organic fertilizers at transplant—which could be possible by means of a fertilizer tank mounted on the direct transplanting machine—and fertigation with organic material, may result in a better stratification of fertilizers even in no-till conditions, allowing the reduction of the exposure of organic fertilisers to oxidation conditions, while increasing their efficiency.

Moreover, the trade-off between GHG mitigation and the crop productivity has to be taken into account, evaluating the crop yield in the three cropping systems [63].

5. Conclusions

The ORG+ system registered a tendency of higher CO_2 emissions and lower N_2O emissions respect to INT and ORG systems. The lower N_2O emissions were probably related to the low N rate supplied in ORG+, while the higher CO_2 emissions could have been due to the higher supply of organic material with organic fertilizer and to the higher autotrophic respiration due to living mulch. No differences among the three systems were observed concerning CH_4 emissions. Based on our results, the organic conservation system did not show a clear tendency towards mitigating soil GHG emissions in vegetable rotation in a Mediterranean environment.

Further soil GHG monitoring campaigns are needed to compare the three systems in the long term. Moreover, other studies will be needed to assess the overall sustainability of the three cropping systems from an agronomic, economic and environmental (e.g., life cycle assessment) point of view.

Supplementary Materials: The following are available online at http://www.mdpi.com/2073-4395/9/8/446/s1, **Figure S1**: (a) Relationship between WFPS and N_2O daily flux; (b) relationship between soil temperature and CO_2 daily flux., **Table S1**: Bibliographic references for C:N of each crop in the green manures., **Table S2**: Estimated values of C:N for the green manure mixtures (ORG) during the field experiment period.

Author Contributions: Conceptualization, S.B., D.A. and C.F.; data curation, S.B. and I.V.; formal analysis, I.V.; methodology, S.B. and D.A.; project administration, C.F.; software, I.V.; supervision, S.B. and C.F.; validation, G.R.; visualization, S.B. and I.V.; writing—original draft, S.B. and I.V.; writing—review and editing, D.A. and G.R..

Funding: This research was carried out within the project SMOCA "Smart Management of Organic Conservation Agriculture" (http://smoca.agr.unipi.it/) funded by the Italian Ministry of University and Research (MIUR) within the program FIRB-2013 (*Future in Research*), MIUR-FIRB13 (project number: RBFR13L8J6).

Acknowledgments: The authors would like to acknowledge Enrico Bonari for helpful discussions and comments and Cristiano Tozzini, Fabio Taccini, Jonatha Trabucco and Tommaso Bambini of the Institute of Life Sciences, Scuola Superiore Sant'Anna, for facilitating our field work. We thank the staff at the "Enrico Avanzi" Centre for Agro-Environmental Research of the University of Pisa, who managed the field trials and provided technical support throughout.

Conflicts of Interest: The authors declare no conflicts of interest.

References

1. Cook, S.K.; Collier, R.; Clarke, J.; Lillywhite, R. Contribution of integrated farm management (IFM) to Defra objectives. *Asp. Appl. Biol.* **2009**, *93*, 131–138.
2. Morris, C.; Winter, M. Integrated farming systems: The third way for European agriculture? *Land Use Policy* **1999**, *16*, 193–205. [CrossRef]
3. Peigné, J.; Ball, B.C.; Roger-Estrade, J.; David, C. Is conservation tillage suitable for organic farming? A review. *Soil Use Manag.* **2007**, *23*, 129–144. [CrossRef]

4. Peigné, J.; Casagrande, M.; Payet, V.; David, C.; Sans, F.X.; Blanco-Moreno, J.M.; Cooper, J.; Gascoyne, K.; Antichi, D.; Bàrberi, P.; et al. How organic farmers practice conservation agriculture in Europe. *Renew. Agric. Food Syst.* **2015**, *31*, 72–85. [CrossRef]

5. González-Sánchez, E.J.; Ordóñez-Fernández, R.; Carbonell-Bojollo, R.; Veroz-González, O.; Gil-Ribes, J.A. Meta-analysis on atmospheric carbon capture in Spain through the use of conservation agriculture. *Soil Tillage Res.* **2012**, *122*, 52–60. [CrossRef]

6. Hobbs, P.; Sayre, K.; Gupta, R. The role of conservation agriculture in sustainable agriculture. *Philos. Trans. R. Soc. Lond. B Biol. Sci.* **2008**, *363*, 543–555. [CrossRef] [PubMed]

7. Holland, J.M. The environmental consequences of adopting conservation tillage in Europe: Reviewing the evidence. *Agric. Ecosyst. Environ.* **2004**, *103*, 1–25. [CrossRef]

8. Sans, F.X.; Berner, A.; Armengot, L.; Mäder, P. Tillage effects on weed communities in an organic winter wheat-sunflower-spelt cropping sequence. *Weed Res.* **2011**, *51*, 413–421. [CrossRef]

9. IFOAM. *The IFOAM Norms for Organic Production and Processing*; IFOAM: Bonn, Germany, 2012.

10. Shirtliffe, S.J.; Johnson, E.N. Progress towards no-till organic weed control in western Canada. *Renew. Agric. Food Syst.* **2012**, *27*, 60–67. [CrossRef]

11. Snyder, C.S.; Bruulsema, T.W.; Jensen, T.L.; Fixen, P.E. Review of greenhouse gas emissions from crop production systems and fertilizer management effects. *Agric. Ecosyst. Environ.* **2009**, *133*, 247–266. [CrossRef]

12. Powlson, D.S.; Whitmore, A.P.; Goulding, K.W.T. Soil carbon sequestration to mitigate climate change: A critical re-examination to identify the true and the false. *Eur. J. Soil Sci.* **2011**, *62*, 42–55. [CrossRef]

13. Palm, C.; Blanco-canqui, H.; Declerck, F.; Gatere, L.; Grace, P. Conservation agriculture and ecosystem services: An overview. *Agric. Ecosyst. Environ.* **2014**, *187*, 87–105. [CrossRef]

14. Aguilera, E.; Lassaletta, L.; Gattinger, A.; Gimeno, B.S. Managing soil carbon for climate change mitigation and adaptation in Mediterranean cropping systems: A meta-analysis. *Agric. Ecosyst. Environ.* **2013**, *168*, 25–36. [CrossRef]

15. Thorup-Kristensen, K.; Magid, J.; Jensen, L.S. Catch crops and green manures as biological tools in nitrogen management in temperate zones. *Adv. Agron.* **2003**, *79*, 227–302.

16. Kaye, J.P.; Quemada, M. Using cover crops to mitigate and adapt to climate change. A review. *Agron. Sustain. Dev.* **2017**, *37*, 4. [CrossRef]

17. Muhammad, I.; Sainju, U.M.; Zhao, F.; Khan, A.; Ghimire, R.; Fu, X. Regulation of soil CO_2 and N_2O emissions by cover crops: A meta-analysis. *Soil Tillage Res.* **2019**, *192*, 103–112. [CrossRef]

18. Baggs, E.M.; Stevenson, M.; Pihlatie, M.; Regar, A.; Cook, H.; Cadisch, G. Nitrous oxide emissions following application of residues and fertiliser under zero and conventional tillage. *Plant Soil* **2003**, *254*, 361–370. [CrossRef]

19. Basche, A.D.; Miguez, F.E.; Kaspar, T.C.; Castellano, M.J. Do cover crops increase or decrease nitrous oxide emissions? A meta-analysis. *J. Soil Water Conserv.* **2014**, *69*, 471–482. [CrossRef]

20. Turner, P.A.; Baker, J.M.; Griffis, T.J.; Venterea, R.T. Impact of Kura Clover Living Mulch on Nitrous Oxide Emissions in a Corn–Soybean System. *J. Environ. Qual.* **2016**, *45*, 1782. [CrossRef]

21. Six, J.; Ogle, S.M.; Breidt, F.J.; Conant, R.T.; Mosiers, A.R.; Paustian, K. The potential to mitigate global warming with no-tillage management is only realized when practised in the long term. *Glob. Chang. Biol.* **2004**, *10*, 155–160. [CrossRef]

22. Hutsch, B.W. Methane oxidation in non-flooded soils as affected by crop production—Invited paper. *Eur. J. Agron.* **2001**, *14*, 237–260. [CrossRef]

23. Ussiri, D.A.N.; Lal, R.; Jarecki, M.K. Nitrous oxide and methane emissions from long-term tillage under a continuous corn cropping system in Ohio. *Soil Tillage Res.* **2009**, *104*, 247–255. [CrossRef]

24. Soil Survey Staff. *Keys to Soil Taxonomy*, 12th ed.; USDA-Natural Resources Conservation Service: Washington, DC, USA, 2014.

25. Livingston, G.P.; Hutchinson, G.L. Enclosure-based measurement of trace gas exchange: Applications and sources of error. In *Methods in Ecology. Biogenic Trace Gases: Measuring Emissions from Soil and Water*; Matson, P.A., Harriss, R.C., Eds.; Blackwell Science: Malden, MA, USA, 1995; pp. 14–51.

26. Bosco, S.; Volpi, I.; Nassi o Di Nasso, N.; Triana, F.; Roncucci, N.; Tozzini, C.; Villani, R.; Laville, P.; Neri, S.; Mattei, F.; et al. LIFE+IPNOA mobile prototype for the monitoring of soil N_2O emissions from arable crops: First-year results on durum wheat. *IJA* **2015**, *10*, 124. [CrossRef]

27. Laville, P.; Neri, S.; Continanza, D.; Vero, L.F.; Bosco, S.; Virgili, G. Cross-Validation of a Mobile N_2O Flux Prototype (IPNOA) Using Micrometeorological and Chamber Methods. *J. Energy Power Eng.* **2015**, *9*, 375–380.

28. Laville, P.; Bosco, S.; Volpi, I.; Virgili, G.; Neri, S.; Continanza, D.; Bonari, E. Temporal integration of soil N_2O fluxes: Validation of IPNOA station automatic chamber prototype. *Environ. Monit. Assess.* **2017**, *189*, 485. [CrossRef] [PubMed]

29. R Core Team. *R: A Language and Environment for Statistical Computing*; R Foundation for Statistical Computing: Vienna, Austria, 2018; Available online: https://www.R-project.org/ (accessed on 28 February 2018).

30. Bates, D.; Maechler, M.; Bolker, B.; Walker, S.; Christensen, R.H.B.; Singmann, H.; Dai, B. Linear mixed-effects models using Eigen and S4. R package version 1.1-7. 2014. Available online: http://CRAN.R-project.org/package=lme4 (accessed on 12 March 2018).

31. Myhre, G.; Shindell, D.; Bréon, F.-M.; Collins, W.; Fuglestvedt, J.; Huang, J.; Koch, D.; Lamarque, J.-F.; Lee, D.; Mendoza, B.; et al. *Anthropogenic and Natural Radiative Forcing: In Climate Change 2013: The Physical Science Basis. Contribution of Working Group I to the Fifth Assessment Report of the Intergovernmental Panel on Climate Chang*; Cambridge University Press: Cambridge, UK; New York, NY, USA, 2013; pp. 659–740.

32. Skiba, U.; Smith, K.A. The control of nitrous oxide emissions from agricultural and natural soils. *Chemosphere-Glob. Chang. Sci.* **2000**, *2*, 379–386. [CrossRef]

33. Liu, C.; Wang, K.; Meng, S.; Zheng, X.; Zhou, Z.; Han, S.; Chen, D.; Yang, Z. Effects of irrigation, fertilization and crop straw management on nitrous oxide and nitric oxide emissions from a wheat-maize rotation field in northern China. *Agric. Ecosyst. Environ.* **2011**, *140*, 226–233. [CrossRef]

34. Flessa, H.; Ruser, R.; Dörsch, P.; Kamp, T.; Jimenez, M.A.; Munch, J.C.; Beese, F. Integrated evaluation of greenhouse gas emissions (CO_2, CH_4, N_2O) from two farming systems in southern Germany. *Agric. Ecosyst. Environ.* **2002**, *91*, 175–189. [CrossRef]

35. Snyder, C.S.; Davidson, E.A.; Smith, P.; Venterea, R.T. Agriculture: Sustainable crop and animal production to help mitigate nitrous oxide emissions. *Curr. Opin. Environ.Sustain.* **2014**, *9*, 46–54. [CrossRef]

36. Volpi, I.; Laville, P.; Bonari, E.; Nassi O Di Nasso, N.; Bosco, S. Nitrous oxide mitigation potential of reduced tillage and N input in durum wheat in the Mediterranean. *Nutr. Cycl. Agroecosyst.* **2018**, *111*, 189–201. [CrossRef]

37. De Rosa, D.; Rowlings, D.W.; Biala, J.; Scheer, C.; Basso, B.; Grace, P.R. N_2O and CO_2 emissions following repeated application of organic and mineral N fertiliser from a vegetable crop rotation. *Sci. Total Environ.* **2018**, *637–638*, 813–824. [CrossRef] [PubMed]

38. Thangarajan, R.; Bolan, N.S.; Tian, G.; Naidu, R.; Kunhikrishnan, A. Role of organic amendment application on greenhouse gas emission from soil. *Sci. Total Environ.* **2013**, *465*, 72–96. [CrossRef] [PubMed]

39. Pelster, D.E.; Chantigny, M.H.; Rochette, P.; Angers, D.A.; Rieux, C.; Vanasse, A. Nitrous Oxide Emissions Respond Differently to Mineral and Organic Nitrogen Sources in Contrasting Soil Types. *J. Environ.Qual.* **2012**, *41*, 427–435. [CrossRef] [PubMed]

40. Webb, J.; Pain, B.; Bittman, S.; Morgan, J. The impacts of manure application methods on emissions of ammonia, nitrous oxide and on crop response—A review. *Agric. Ecosyst. Environ.* **2010**, *137*, 39–46. [CrossRef]

41. Heller, H.; Bar-Tal, A.; Tamir, G.; Bloom, P.; Venterea, R.T.; Chen, D.; Zhang, Y.; Clapp, C.E.; Fine, P. Effects of Manure and Cultivation on Carbon Dioxide and Nitrous Oxide Emissions from a Corn Field under Mediterranean Conditions. *J. Environ. Qual.* **2010**, *39*, 437. [CrossRef] [PubMed]

42. Ambus, P.; Jensen, E.; Robertson, G. Nitrous oxide and water mediated N-losses from agricultural soil: Influence of crop residue particle size, quality and placement. *Phyton (Austria)* **2001**, *41*, 7–15.

43. Chen, H.; Li, X.; Hu, F.; Shi, W. Soil nitrous oxide emissions following crop residue addition: A meta-analysis. *Glob. Chang Biol.* **2013**, *19*, 2956–2964. [CrossRef]

44. Pugesgaard, S.; Petersen, S.O.; Chirinda, N.; Olesen, J.E. Crop residues as driver for N_2O emissions from a sandy loam soil. *Agric. Meteorol.* **2017**, *233*, 45–54. [CrossRef]

45. Flessa, H.; Ruser, R.; Schilling, R.; Loftfield, N.; Munch, J.C.; Kaiser, E.A.; Beese, F. N_2O and CH_4 fluxes in potato fields: Automated measurement, management effects and temporal variation. *Geoderma* **2002**, *105*, 307–325. [CrossRef]

46. Abdalla, M.; Osborne, B.; Lanigan, G.; Forristal, D.; Williams, M.; Smith, P.; Jones, M.B. Conservation tillage systems: A review of its consequences for greenhouse gas emissions. *Soil Use Manag.* **2013**, *29*, 199–209. [CrossRef]

47. Skinner, C.; Gattinger, A.; Muller, A.; Mäder, P.; Fliebach, A.; Stolze, M.; Ruser, R.; Niggli, U. Greenhouse gas fluxes from agricultural soils under organic and non-organic management—A global meta-analysis. *Sci. Total Environ.* **2014**, *468–469*, 553–563. [CrossRef] [PubMed]

48. Pittelkow, C.M.; Linquist, B.A.; Lundy, M.E.; Liang, X.; Van Groenigen, K.J.; Lee, J.; Van Gestel, N.; Six, J.; Venterea, R.T.; Van Kessel, C. When does no-till yield more? A global meta-analysis. *Field Crop. Res.* **2015**, *183*, 156–168. [CrossRef]

49. Sanz-Cobena, A.; García-Marco, S.; Quemada, M.; Gabriel, J.L.; Almendros, P.; Vallejo, A. Do cover crops enhance N_2O, CO_2 or CH_4 emissions from soil in Mediterranean arable systems? *Sci. Total Environ.* **2014**, *466–467*, 164–174. [CrossRef] [PubMed]

50. Guardia, G.; Tellez-Rio, A.; García-Marco, S.; Martin-Lammerding, D.; Tenorio, J.L.; Ibáñez, M.Á.; Vallejo, A. Effect of tillage and crop (cereal versus legume) on greenhouse gas emissions and Global Warming Potential in a non-irrigated Mediterranean field. *Agric. Ecosyst. Environ.* **2016**, *221*, 187–197. [CrossRef]

51. Serrano-Silva, N.; Sarria-Guzmán, Y.; Dendooven, L.; Luna-Guido, M. Methanogenesis and Methanotrophy in Soil: A Review. *Pedosphere* **2014**, *24*, 291–307. [CrossRef]

52. Forte, A.; Fagnano, M.; Fierro, A. Potential role of compost and green manure amendment to mitigate soil GHGs emissions in Mediterranean drip irrigated maize production systems. *J. Environ. Manag.* **2017**, *192*, 68–78. [CrossRef] [PubMed]

53. Lloyd, J.; Taylor, J.A. On the Temperature Dependence of Soil Respiration. *Funct. Ecol.* **1994**, *8*, 315–323. [CrossRef]

54. Davidson, E.A.; Janssens, I.A.; Lou, Y. On the variability of respiration in terrestrial ecosystems: Moving beyond Q10. *Glob. Chang. Biol.* **2006**, *12*, 154–164. [CrossRef]

55. Lai, R.; Arca, P.; Lagomarsino, A.; Cappai, C.; Seddaiu, G.; Demurtas, C.E.; Roggero, P.P. Manure fertilization increases soil respiration and creates a negative carbon budget in a Mediterranean maize (*Zea mays* L.)-based cropping system. *Catena* **2017**, *151*, 202–212. [CrossRef]

56. Almagro, M.; López, J.; Querejeta, J.I.; Martínez-Mena, M. Temperature dependence of soil CO_2 efflux is strongly modulated by seasonal patterns of moisture availability in a Mediterranean ecosystem. *Soil Biol. Biochem.* **2009**, *41*, 594–605. [CrossRef]

57. Mancinelli, R.; Marinari, S.; Di Felice, V.; Savin, M.C.; Campiglia, E. Soil property, CO_2 emission and aridity index as agroecological indicators to assess the mineralization of cover crop green manure in a Mediterranean environment. *Ecol. Indic.* **2013**, *34*, 31–40. [CrossRef]

58. Morell, F.J.; Álvaro-Fuentes, J.; Lampurlanés, J.; Cantero-Martínez, C. Soil CO_2 fluxes following tillage and rainfall events in a semiarid Mediterranean agroecosystem: Effects of tillage systems and nitrogen fertilization. *Agric. Ecosyst. Environ.* **2010**, *139*, 167–173. [CrossRef]

59. Abdalla, M.; Hastings, A.; Helmy, M.; Prescher, A.; Osborne, B.; Lanigan, G.; Forristal, D.; Killi, D.; Maratha, P.; Williams, M.; et al. Assessing the combined use of reduced tillage and cover crops for mitigating greenhouse gas emissions from arable ecosystem. *Geoderma* **2014**, *223–225*, 9–20. [CrossRef]

60. Negassa, W.; Price, R.F.; Basir, A.; Snapp, S.S.; Kravchenko, A. Cover crop and tillage systems effect on soil CO_2 and N_2O fluxes in contrasting topographic positions. *Soil Tillage Res.* **2015**, *154*, 64–74. [CrossRef]

61. Chirinda, N.; Carter, M.S.; Albert, K.R.; Ambus, P.; Olesen, J.E.; Porter, J.R.; Petersen, S.O. Emissions of nitrous oxide from arable organic and conventional cropping systems on two soil types. *Agric. Ecosyst. Environ.* **2010**, *136*, 199–208. [CrossRef]

62. Van Kessel, C.; Venterea, R.; Six, J.; Adviento-Borbe, M.A.; Linquist, B.; van Groenigen, K.J. Climate, duration, and N placement determine N_2O emissions in reduced tillage systems: A meta-analysis. *Glob. Chang. Biol.* **2013**, *19*, 33–44. [CrossRef] [PubMed]

63. Antichi, D.; Sbrana, M.; Martelloni, L.; Abou Chehade, L.; Fontanelli, M.; Raffaelli, M.; Mazzoncini, M.; Peruzzi, A.; Frasconi, C. Agronomic performances of organic field vegetables managed with conservation agriculture techniques: A study from Central Italy. **2019**, Manuscript in preparation.

Article

Evaluation of the Agronomic Performance of Organic Processing Tomato as Affected by Different Cover Crop Residues Management

Lara Abou Chehade [1,*], Daniele Antichi [1], Luisa Martelloni [1], Christian Frasconi [1], Massimo Sbrana [2], Marco Mazzoncini [1] and Andrea Peruzzi [1]

[1]   Department of Agriculture, Food and Environment, University of Pisa, via del Borghetto 80, 56124 Pisa, Italy
[2]   Center for Agri-environmental Research "Enrico Avanzi", University of Pisa, via Vecchia di Marina 6, San Piero a Grado, 56122 Pisa, Italy
*   Correspondence: lara.abouchehade@agr.unipi.it

Received: 11 July 2019; Accepted: 28 August 2019; Published: 1 September 2019

Abstract: No-till practices reduce soil erosion, conserve soil organic carbon, and enhance soil fertility. Yet, many factors could limit their adoption in organic farming. The present study investigated the effects of tillage and cover cropping on weed biomass, plant growth, yield, and fruit quality of an organic processing tomato (*Solanum lycopersicon* L. var. Elba F1) over two seasons (2015–2017). We compared systems where processing tomato was transplanted on i) tilled soil following or not a winter cover crop (*Trifolium squarrosum* L.) and with/without a biodegradable plastic mulch; and ii) no-till where clover was used, after rolling and flaming, as dead mulch. Tomato in no-till suffered from high weed competition and low soil nitrogen availability leading to lower plant growth, N uptake, and yield components with respect to tilled systems. The total yield in no-till declined to 6.8 and 18.3 t ha^{-1} in 2016 and 2017, respectively, with at least a 65% decrease compared to tilled clover-based systems. No evidence of growth-limiting soil compaction was noticed but a slightly higher soil resistance was in the no-till topsoil. Tillage and cover crop residues did not significantly change tomato quality (pH, total soluble solids, firmness). The incorporation of clover as green manure was generally more advantageous over no-till. This was partly due to the low performance of the cover crop where improvement may limit the obstacles (i.e., N supply and weed infestation) and enable the implementation of no-till in organic vegetable systems.

Keywords: no-till; green manure; dead mulch; biodegradable plastic mulch; organic farming; conservation agriculture; tomato

---

1. Introduction

According to recent statistics, land managed under organic farming regulations in Europe has increased by almost 75% in the last decade [1]. Consumer demand for environmental sustainability as well as safety and food quality concerns continue to drive the organic industry and to encourage farmers to convert their agricultural systems to organic farming. However, organic producers rely primarily on intensive and frequent tillage for weed management, organic fertilizers and residue incorporation, and seedbed preparation [2], in a way that sometimes violates the objective of organic farming to sustain soil health. Intensive tillage reduces soil quality, facilitates erosion through the destruction of soil structure, increases loss of topsoil organic matter, and decreases soil biological activity and biodiversity [3]. No-tillage systems were developed a few decades ago in conventional agriculture to mitigate these problems and to provide economic savings by eliminating tillage and excessive traffic on fields [3–5]. Benefits to soil fertility and other ecological services (i.e., weed and pest suppression, nutrient cycling) are provided by cultivation of cover crops in rotation with cash crops as well [6]. Using legume species as cover crops also provides additional N fixed from the atmosphere into the agro-ecosystem thus improving N nutrition of the cash crop and increasing soil nitrogen organic pool [7].

Recently, researchers have been increasingly investigating cover-crop-based no-till (NT) as a sustainable practice to eliminate the reliance on mechanical tillage and maximize the benefits of cover crops and resource use efficiency in organic farming [6,8]. In these systems, cover crops are terminated without incorporating residues into the soil, thus leaving a thick mulch into which the subsequent cash crop is planted. This requires the necessity to produce large cover crop biomass as well as a good management of their residues to provide maximum weed suppression and nutrients adjustments, e.g., reduce immobilization, enhance N release and synchronization with plant needs [9]. Weed management and nutrient availability are two factors known to challenge the performance of crops in organic no-till production. In such systems, weeds tend to increase with higher seedling recruitment in the upper soil layers and large infestations of perennial weeds [2,10]. Cover crops can reduce weed infestation during their growth and/or by their residues on soil surface making a physical barrier, preventing sunlight reaching the soil surface or through allelopathy [11]. With reduced or absence of tillage, mineralization of soil organic matter can also be slowed down which would make N a limiting factor in these conditions and compromise yield production [12–14].

Italy is the second largest producer of processing tomato after the USA with more than 72,000 ha dedicated to it as of 2018 [15,16]. In this study, we aimed to understand how the transition to no-till would impact the production of tomato and if a mulch of cover crop residues would be able to replace plastic mulch which is costly and difficult to dispose of when the material is not biodegradable. To this end, the following field experiments (2015–2017) compared cover crop-based no-till and conventionally tilled systems for organic processing tomato production under Mediterranean conditions in terms of crop growth, yield, fruit quality, N uptake as well as the changes in soil nitrates, soil compaction, and weed infestation.

2. Materials and Methods

2.1. Field and Treatments Description

The experiments were conducted on certified organic fields at the Center for Agri- Environmental Research "Enrico Avanzi" of the University of Pisa (San Piero a Grado, Pisa, Italy) for two seasons (2015–2017). Seven systems were adopted: squarrose clover (*Trifolium squarrosum* L.) rolled, flamed, and followed by a direct transplantation of tomato (*Solanum lycopersicon* L. var. Elba F1, a processing cultivar that can be used also for fresh consumption) (NT-CC); squarrose clover rolled and flamed, followed by a direct transplantation of tomato and supplemented with weeding interventions, i.e., inter-row mowing (NT-CC-SW); squarrose clover incorporated as green manure (CT-CC); squarrose clover incorporated and the soil covered with black biodegradable plastic mulch (Mater-Bi®) set over the season (CT-CC-PM); fallow conventionally-tilled soil covered with plastic mulch (CT-NC-PM), fallow conventionally-tilled with soil kept bare (CT-NC), and a weedy control left untilled with natural vegetation (NT-NC). The fields were moldboard ploughed and harrowed in 16 November 2015 and 3 October 2016 before the cover crop (*T. squarrosum*) broadcast manual sowing at 50 kg ha^{-1} seeding rate on 17 December 2015 and 35 kg ha^{-1} on 12 October 2016. The sowing densities of the clover differed across the two years because of the different germination rate and of the delayed sowing date in 2015, but they were targeted to the same plant densities (667 plants m^{-2}).

In conventionally tilled (CT) plots, the cover crop was terminated using a rotary hoe at around 15 cm depth. Fallow plots were prepared for transplanting the same way. In these plots, inter-row cultivation was also performed for subsequent weed control. The cover crop in NT treatments was terminated with two passes of a roller-crimper (Eco-roll, Clemens Technologies, Wittlich, Germany) followed by one pass of flaming (MAITO Srl., Arezzo, Italy based on a prototype designed and fully realized at the University of Pisa) to enhance cover crop devitalization [17] on 23 May 2016 and 10 May

2017. In NT plots with supplemental weeding, three inter-row mowing interventions (lawn mower) were done during the early season of tomato growth.

In plots without plastic mulch, tomato seedlings were transplanted on 23 June 2016 and 11 May 2017 at a density of 2.22 plants m^{-2} (0.3 m along the row, 1.5 m between the row) with a commercial vegetable transplanting machine ("Fast" model, Fedele costruzioni Meccaniche, Lanciano, Chieti, Italy) modified at the University of Pisa in order to be properly used both on tilled and untilled soil [18]. Tomato seedlings were instead manually transplanted at the same plant density on plastic mulch systems. The distance between tomato single rows (1.5 m), fine-tuned for the plastic mulch system, was kept the same for all treatments to avoid additional variability that can influence the results and conclusions. Phytosanitary measures followed European organic farming regulations. During the growing seasons, fertigation was done at modest doses providing around 16 kg ha^{-1} N and 32 kg ha^{-1} K$_2$O (VIT-ORG) for all systems alike. The fertilization was meant to avoid K lack during fruit ripening, keeping the N supply at a minimum level (i.e., the amount of N contained in the NK fertilizer) in order to avoid masking the effects of treatments on N availability. The fertigation was practiced twice each year (when at least 70% of plants in all the plots reached the fruit set stage and two weeks later) with a single irrigation intervention early in the morning. Plots were 10 m × 6 m and 10 m × 5 m wide, respectively, in 2016 and 2017 and were distributed in a completely randomized block design over different fields each year. The cover crop at killing dates yielded 2.3 (SD = 0.98) and 3.5 t ha^{-1} (SD = 1.6) of dry biomass and had a N yield of 49.1 and 75.9 kg ha^{-1} in 2016 and 2017, respectively. The soil was a sandy loam in 2015–2016 and a sandy clay loam in 2016–2017. Soil characteristics in each experimental site/year are detailed in Table 1. Weather conditions reported for the last 25 years and during the experiment are also presented in Figure 1.

Table 1. Soil characteristics of the fields where the experiments were carried out.

Characteristic	Measurement Unit	2015–2016	2016–2017
Clay	g 100 g^{-1}	11.67	21.80
Silt	g 100 g^{-1}	18.24	4.70
Sand	g 100 g^{-1}	70.09	73.50
pH		7.89	7.89
EC	µS	48.12	45.23
Total N	g kg^{-1}	1.27	0.76
SOM	g 100 g^{-1}	1.97	1.27
P available	µg 100 g^{-1}	2.43	4.20

The Kjeldahl method was used for total N determination, the Walkley–Black method for soil organic matter (SOM), and the Olsen P test for soil available phosphorus (P) determination. EC = electrical conductivity.

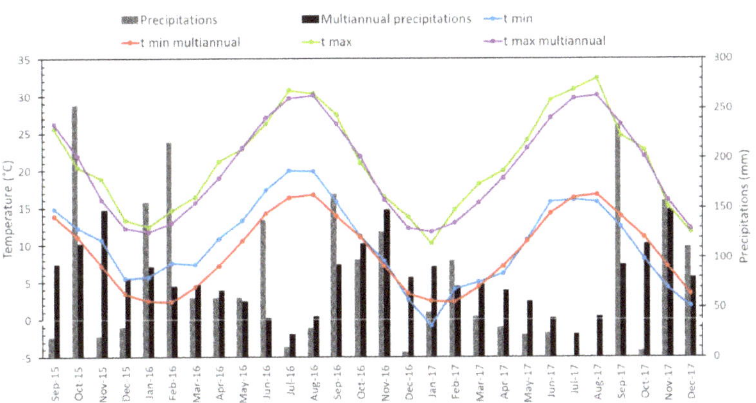

Figure 1. Total monthly precipitations (mm) and the average maximum and minimum temperatures

(°C) during 2015 and 2017 compared to the multiannual average precipitations and temperatures (1993–2017) in San Piero a Grado, Pisa.

2.2. Field Samplings and Measurements

Tomato fruits were harvested from 12 plants of the central row of each plot through the season. The cumulative number of discarded (i.e., diseased, rotten, damaged), green and marketable tomatoes, and their corresponding fresh weights were recorded. Total yield as the sum of the fresh weights of all categories was therefore determined in order to estimate the potential cumulative yield of tomato. The dry matter (DM) content of fruits was obtained by oven-drying a sample at 60 °C until a constant weight was obtained. At the end of the harvest period, tomato residues and weeds were simultaneously collected over two areas of 1 m^2 in each plot. For crop residues, plants were excavated at depths of 25–30 cm and shoots and roots were separated after cleaning roots from soil residues. Plant parts were then oven-dried at 60 °C for dry matter and N content determination [19]. Tomato total N uptake (kg N ha^{-1}) was calculated as:

$$N\ uptake = (a \times b) + (c \times d) + (e \times f) \tag{1}$$

where "a" is tomato yield (kg ha^{-1} of DM), "b" is the N concentration of marketable tomato fruits (g 100 g^{-1} of DM), "c" is the tomato shoot yield (kg ha^{-1} of DM), and "d" is the N concentration of tomato shoot (g 100 g^{-1} of DM), "e" is the root yield (kg ha^{-1} of DM), and "f" is the N concentration of tomato root.

To assess the dynamic status of nitrogen in the soil, the nitrate content was determined [20] every 10–20 days at a depth of 30 cm on a composite sample (2 samples) from each plot, starting at transplantation and continuing during the season. A hand-held electronic cone-tipped penetrometer (Spectrum Field Scout SC-900, Spectrum Technologies Inc., Plainfield, IL, USA) was used to measure soil resistance (KPa) on three different locations in each plot at harvest across a 45 cm soil depth.

2.3. Fruit Quality

Fruit firmness was measured in 2016 using a digital fruit firmness tester (penetrometer) with an 8 mm diameter plunger (TR Turoni srl, Forlì, Italy). The peak force or the maximum force to compress the fruits by 5 mm determined between two parallel plates using an Instron Universal Testing Machine (Model 3343, Norwood, MA, USA) was recorded as an indicator of the firmness of tomato fruits in 2017. A pH meter (Cyberscan pH 110, Eutech instruments, Singapore and Titrator T50, Mettler Toledo, Greifensee, Switzerland) was used to determine the pH. The total soluble sugars (TSS) of the juice was determined by a digital hand-held refractometer (Atago PR32-Palette, Tokyo, Japan) and expressed as °Brix. Vitamin C as the sum of both ascorbic and dehydroascorbic acid was determined in 2017 on fresh tomato as in Zapata and Dufour [21] with some modifications [22] as well as the total phenolic content and the antioxidant activity of tomato [23,24]. Quality measurements were performed on around 10 to 15 red marketable fruits from each treatment.

2.4. Statistical Analyses

General linear mixed-effect models for the analysis of variance (ANOVA) were used using R statistical software and the *lmerTest* package to check for the effects of treatments and years, after verifying the normality and homoscedasticity of errors. In the case of fruit number, data were modelled in a generalized linear mixed-effect model (*lmerTest* package) using Poisson distribution. In all models, treatments (systems) and years were used as fixed factors, and blocks and years as random ones. Data were presented separated by year because of the significant year effect and interaction between year and treatment in most of the cases. Pairwise comparisons for all variables were computed by estimating the 95% confidence interval of the difference between the least squares means (Equation (2)):

$$CI(\text{difference}) = (x_1 - x_2) \pm 1.96 \sqrt{(SE_{x1})^2 + (SE_{x2})^2} \tag{2}$$

where (x_1) is the mean of the first value, (x_2) is the mean of the second value, (SE_{x1}) is the standard error of (x_1), and (SEx_2) is the standard error of (x_2).

If the resulting 95% confidence interval (CI) of the difference between values did not cross the zero value, the null hypothesis that the compared values are similar was rejected.

All data in the manuscript were reported in the original scale as least square means with their corresponding standard errors. Results of all analysis of variance/deviance in terms of *p*-values are presented in Table S1.

3. Results

3.1. Plant Biomass and N Uptake

Plant biomass was influenced by treatments and the growing year. Higher biomass of fruits and shoots were obtained in systems where clover was incorporated into the soil with and without plastic mulch compared to no-till cover crop-based systems in 2016 (Table 2). Only the treatment where clover was incorporated under plastic mulch resulted in higher root biomass that year.

Table 2. Plant dry biomass as affected by tillage and cover crop management.

	2015–2016			2016–2017		
Treatment	Fruits Dry Biomass (g m^{-2})	Shoots Dry Biomass (g m^{-2})	Roots Dry Biomass (g m^{-2})	Fruits Dry Biomass (g m^{-2})	Shoots Dry Biomass (g m^{-2})	Roots Dry Biomass (g m^{-2})
CT-CC	264.4 [a]	181.8 [a]	22.3 [b]	354.2 [b]	279.4 [b]	28.4 [c]
CT-CC-PM	223.0 [b]	183.2 [a]	28.0 [a]	436.6 [a]	372.7 [a]	36.8 [a]
CT-NC	257.4 [a,b]	166.6 [a]	22.6 [b]	355.3 [b]	250.1 [c]	33.1 [b]
CT-NC-PM	240.2 [a,b]	159.0 [a]	16.9 [c]	405.3 [a,b]	293.4 [b]	39.3 [a]
NT-CC-SW	48.4 [c]	31.5 [b]	4.6 [d]	171.5 [c]	137.5 [d]	22.7 [d]
NT-CC	45.9 [c]	26.0 [b]	4.0 [d]	129.8 [d]	101.7 [e]	15.1 [e]
NT-NC	30.5 [c]	24.2 [b]	3.9 [d]	58.1 [e]	53.5 [f]	7.9 [f]
SE	12.9	10.0	1.3	12.9	9.4	1.3

CT-CC: conventionally tilled + cover crop; CT-CC-PM: conventionally tilled + cover crop + plastic mulch; CT-NC: conventionally tilled without cover crop; CT-NC-PM: conventionally tilled without cover crop + plastic mulch; NT-CC-SW: no-till + cover crop + supplemental weeding; NT-CC: no-till + cover crop; NT-NC: no-till without cover crop (weedy control). SE = standard error. Values followed by different letters are significantly different at $p < 0.05$.

In 2017, dry biomass of fruits, shoots, and roots revealed the outperformance of plastic mulch systems over the other systems mainly where clover was incorporated as green manure. The dead mulch had the lowest performance among residue management techniques for all the biomass components. The supplemental weeding over the dead mulch increased fruits', roots', and shoots' dry matter. Generally, plants of all treatments had better performance in that season compared to 2016.

Nitrogen uptake in both seasons followed almost the same trend of the plant biomass which was the main contributor to it (Table 3). Total N uptake in 2016 was higher in conventionally tilled plots over no-till with no significant differences between clover incorporated and clover incorporated in soil covered with plastic mulch. In 2017, N uptake was the lowest in no-till plants and the highest in plastic mulch system with green manure due to the large N uptakes in shoots, roots, and fruits. Differences in nitrogen concentration among treatments in the different plant parts were not statistically significant (data not presented).

Table 3. N uptake by tomato plants as affected by tillage and cover crop management.

	2015–2016				2016–2017			
Treatment	Fruits N Uptake (kg ha^{-1})	Shoots N Uptake (kg ha^{-1})	Roots N Uptake (kg ha^{-1})	Total N Uptake (kg ha^{-1})	Fruits N Uptake (kg ha^{-1})	Shoots N Uptake (kg ha^{-1})	Roots N Uptake (kg ha^{-1})	Total N Uptake (kg ha^{-1})
CT-CC	57.1 [a]	32.2 [a,b]	2.2 [b]	91.5 [ab]	59.1 [b]	44.7 [b]	3.4 [b]	107.2 [c]
CT-CC-PM	55.2 [a]	36.6 [a]	3.3 [a]	95.1 [a]	75.1 [a]	75.1 [a]	4.4 [a]	154.6 [a]
CT-NC	52.3 [a]	29.2 [b]	3.1 [a]	84.7 [b]	56.3 [b]	45.3 [b]	3.7 [a,b]	105.4 [c]
CT-NC-PM	54.2 [a]	36.0 [a]	2.1 [b]	92.4 [a,b]	71.7 [a]	45.6 [b]	4.3 [a]	121.6 [b]
NT-CC-SW	10.8 [b]	5.7 [c]	0.5 [c]	17.1 [c]	27.1 [c]	25.1 [c]	2.3 [c]	54.6 [d]
NT-CC	10.1 [b]	4.4 [c]	0.6 [c]	15.1 [c]	19.6 [c]	18.3 [d]	1.3 [d]	39.2 [e]
NT-NC	7.5 [b]	4.6 [c]	0.5 [c]	11.8 [c]	7.6 [d]	6.5 [e]	0.7 [d]	14.8 [f]
SE	3.3	1.8	0.3	3.2	3.3	2.5	0.3	4.7

Values followed by different letters are significantly different at $p < 0.05$.

3.2. Yield Components and Fruit Quality

Treatments and growing season both had effects on yield components. Irrespective of residues management and the presence of plastic mulch, higher total fruit number was obtained in conventionally tilled systems with respect to no-till in 2016 due to the higher number of red fruits (Table 4). However, the production of marketable and unmarketable fruits depended on the treatment adopted. The CT-NC produced the highest marketable fruits and had the lowest proportion of unmarketable fruits number among CT systems. Marketable fruits in that year were lower than 2017 due to the presence of disease and physiological disorder incidences. In 2017, systems where plastic mulch was preceded with clover as green manure produced the highest number of fruits due to the higher production of red marketable (similar to CT-NC-PM) and unmarketable tomatoes alike and resulted in the highest fresh yield for each type, compared to the other systems. The proportion of discarded fruits of the whole fruit production in 2017, however, was not affected by tillage and cover crop presence. All conventionally tilled systems especially where green manure was present produced more green fruits than no-till in that year.

Table 4. Number of tomato fruits obtained in each system as affected by tillage and cover crop management.

	2015–2016				2016–2017			
Treatment	Marketable Fruits (No m^{-2})	Unmarketable Fruits (No m^{-2})	Green Fruits (No m^{-2})	Total Fruits (No m^{-2})	Marketable Fruits (No m^{-2})	Unmarketable Fruits (No m^{-2})	Green Fruits (No m^{-2})	Total Fruits (No m^{-2})
CT-CC	19.0 ± 3.1 [b]	61.3 ± 4.5 [a]	1.6 ± 0.8	82.3 ± 5.2 [a]	59.3 ± 4.4 [b]	52.0 ± 4.2 [b]	6.7 ± 1.5 [a,b]	118.0 ± 6.3 [c]
CT-CC-PM	10.2 ± 2.1 [c]	73.0 ± 5.0 [a]	1.6 ± 0.8	85.0 ± 5.3 [a]	97.7 ± 5.7 [a]	78.0 ± 5.1 [a]	8.3 ± 1.7 [a]	184.0 ± 7.8 [a]
CT-NC	34.5 ± 4.8 [a]	37.7 ± 3.6 [b]	3.3 ± 1.1	76.0 ± 5.0 [a]	48.7 ± 4.0 [b]	41.3 ± 3.7 [b,c]	3.7 ± 1.1 [b,c]	93.7 ± 5.6 [d]
CT-NC-PM	15.4 ± 2.7 [b,c]	62.6 ± 4.6 [a]	3.0 ± 1.0	81.3 ± 5.2 [a]	103.7 ± 5.9 [a]	39.3 ± 3.6 [c,d]	5.0 ± 1.3 [a,b]	147.7 ± 7.0 [b]
NT-CC-SW	4.6 ± 1.3 [d]	11.6 ± 2.0 [c]	2.3 ± 0.9	18.7 ± 2.5 [b]	38.0 ± 3.5 [c]	42.7 ± 3.8 [b,c]	1.3 ± 0.7 [c]	82.0 ± 5.2 [d]
NT-CC	2.9 ± 1.0 [d]	11.6 ± 2.0 [c]	2.0 ± 0.8	16.7 ± 2.3 [b]	33.3 ± 3.3 [c]	30.3 ± 3.2 [d]	1.3 ± 0.7 [c]	65.0 ± 4.6 [e]
NT-NC	5.2 ± 1.4 [d]	6.3 ± 1.4 [c]	1.3 ± 0.7	13.0 ± 2.1 [b]	12.3 ± 2.0 [d]	6.3 ± 1.4 [e]	1.3 ± 0.7 [c]	20.0 ± 2.6 [f]

Values followed by different letters are significantly different at $p < 0.05$.

Therefore, total yield (Table 5) in both years was drastically reduced under no-till-dead mulch conditions, at least 85% in 2016 and 66% in 2017 compared with incorporated clover, with higher productivity where a supplemental weeding was performed in 2017. However, the effect of the different treatments on yield depended on the season. In 2016, production under plastic mulch conditions was similar to tilled systems without cover crop and kept bare during the season (CT-NC). The highest production was achieved where clover was turned as green manure without a plastic mulch (CT-CC) and this was due to the high number and singular weight of tomato fruits. In 2017, the total productivity reached its highest value (60–70 t ha^{-1}) in plastic mulch systems. Squarrose clover incorporated and covered with plastic mulch was obviously the best performing among the different residues management systems. Despite these results, the system where clover was incorporated without plastic mulch seemed to be more stable than the other systems; in 2017, all systems except CT-CC and the weedy control showed an increase in their production.

Table 5. Tomato yield obtained in each system as affected by tillage and cover crop management.

	2015–2016				2016–2017			
	Fresh Yield (kg m^{-2})			Total Yield (t ha^{-1})	Fresh Yield (kg m^{-2})			Total Yield (t ha^{-1})
Treatment	Marketable	Unmarketable	Green		Marketable	Unmarketable	Green	
CT-CC	1.1 [b,c]	3.3 [a]	0.15 [b]	46.7 [a]	3.8 [b]	1.2 [b]	0.38 [a]	53.7 [c]
CT-CC-PM	0.9 [c]	2.9 [a]	0.25 [a]	39.7 [b]	5.2 [a]	1.4 [a]	0.37 [a]	69.9 [a]
CT-NC	2.5 [a]	1.1 [c]	0.12 [b]	37.3 [b]	4.0 [b]	1.1 [b]	0.22 [b]	52.9 [c]
CT-NC-PM	1.3 [b]	2.3 [b]	0.30 [a]	39.6 [b]	4.9 [a]	1.2 [b]	0.18 [b]	62.9 [b]
NT-CC-SW	0.3 [d]	0.4 [d]	0.13 [b]	7.7 [c]	1.9 [c]	0.5 [c]	0.06 [c]	24.4 [d]
NT-CC	0.2 [d]	0.4 [d]	0.10 [b,c]	6.8 [c]	1.4 [c]	0.4 [d]	0.05 [c]	18.3 [e]
NT-NC	0.2 [d]	0.2 [d]	0.06 [c]	4.6 [c]	0.5 [d]	0.1 [e]	0.03 [c]	7.0 [f]
SE	0.1	0.1	0.02	1.9	0.3	0.03	0.02	2.4

Values followed by different letters are significantly different at $p < 0.05$.

Regarding fruit quality, firmness is a mechanical property relevant for both processing and fresh tomatoes. It defines the susceptibility of the fruits to mechanical damage during harvest and transportation as well as to environmental ones like drought and temperature changes. Therefore, plants with higher firmness are less prone to qualitative and quantitative losses and have a longer shelf life [25]. Firmer fruits are preferred for processing purposes to maintain the form and integrity of fruits during transformation. Fruit firmness was the same in all treatments in 2015 and tended to be lower in no-till systems in 2017. The TSS and pH values did not show statistically significant differences among the systems in both years (Table 6). Both factors are important for the final yield, energy saving, and conservation of tomato. Regarding the nutraceutical quality measured only in 2017, vitamin C content increased by at least 32% in plants grown over the dead mulch having 31 mg 100g^{-1} FW. Vitamin C and polyphenols are reported to be the major antioxidant hydrosoluble components in tomato and an increase in their content would be an added value for fresh and processing markets where losses during transformation may occur. In our case, total phenols and the antioxidant activity were not influenced by different tillage and cover crop residues management.

Table 6. Marketable fruit basic and nutraceutical characteristics from each of the systems in comparison.

	2015–2016			2016–2017					
Treatment	Firmness (N)	pH	TSS (°Bx)	Firmness (N)	pH	TSS (°Bx)	Vitamin C (mg 100 g^{-1} FW)	Total Phenols (mg GAE 100 g^{-1} FW)	Antioxidant Activity (mg Trolox 100 g^{-1} FW)
CT-CC	31.3	4.23	4.4	8.5 [a,b,c]	4.58	4.8	20.9 [b]	56.0	65.7
CT-CC-PM	26.5	4.29	4.4	8.2 [b,c]	4.56	5.8	21.4 [b]	67.2	93.2
CT-NC	30.2	4.32	5.8	9.8 [a]	4.54	5.2	23.3 [b]	66.9	81.4
CT-NC-PM	29.8	4.37	4.9	9.3 [a,b]	4.54	4.7	21.2 [b]	56.7	66.7
NT-CC-SW	30.0	4.27	6.4	7.2 [c]	4.67	5.7	26.6 [a,b]	66.7	80.8
NT-CC	27.9	4.19	5.9	6.3 [c,d]	4.52	5.8	30.8 [a]	61.1	79.9
NT-NC	28.8	4.20	6.3	5.7 [d]	4.52	5.3	32.8 [a]	67.2	88.4
SE	2.1	0.05	0.52	0.5	0.07	0.7	2.0	7.4	16.1

Values followed by different letters are significantly different at $p < 0.05$.

3.3. Weed Biomass and Soil Characteristics

Weed biomass at harvest of 2016 was the highest in no-till systems similarly to the weedy control (Table 7), whereas in 2017 the dead mulch succeeded to decrease weed biomass although not at the level of conventionally tilled systems. No effect of supplemental mowing over the dead mulch was seen at harvest time.

Table 7. Weed biomass measured in each system at harvest.

	Weed Biomass (g DW m^{-2})	
Treatment	**2015–2016**	**2016–2017**
CT-CC	67.0 [a,b]	62.2 [c]
CT-CC-PM *	36.9 [b]	28.8 [c]
CT-NC	44.5 [b]	66.5 [c]
CT-NC-PM *	41.7 [b]	30.3 [c]
NT-CC-SW	97.1 [a]	192.7 [b]
NT-CC	110.7 [a]	213.1 [b]
NT-NC	105.1 [a]	343.2 [a]
SE	±16.7	±13.7

* Weed biomass measured on the remaining bare soil of the 1 m^2 area assessed. Values followed by different letters are significantly different at $p < 0.05$.

Soil moisture in 2016 did not show statistical differences among treatments throughout the season, although a trend for higher moisture content under the plastic mulch compared to bare and dead mulch soil at the top 10 cm of the soil was confirmed statistically only in mid-season. In early season 2017, almost all conventionally tilled plots had higher moisture content than no-till systems to a depth of 20 cm, both with and without the dead mulch.

Almost 45 days after cover crop incorporation in 2016 (7 July), soil nitrates content was the highest where clover was incorporated and covered with plastic mulch (CT-CC-PM). Lower NO_3^- were found in soil of plastic mulch without cover crop (CT-NC-PM) and the system where clover was incorporated (CT-CC), while no significant mineralization was seen on dead mulch (NT-CC) (Figure 2). Almost 65 days after clover incorporation/soil preparation, soil nitrates increased in all tilled systems, having a higher nitrates concentration compared to dead mulch. N mineralization in plastic mulch with tilled clover reached a peak after 90 days of clover incorporation (20 August). Nitrogen mineralization continued till 4 months after clover incorporation (20 September), where soil the nitrates content was the highest in plastic mulch systems without significant effect of the green manure. In 2017, after almost 10 days of cover crop incorporation (22 May), nitrogen release started. Nitrates concentration was the highest in plastic mulch with clover (CT-CC-PM) similar to the first season, followed by green manure without plastic mulch and being almost double the lowest concentration found in dead mulch soil. Significant mineralization of green manure clover on bare soil (CT-CC) was detected 24 days after cover crop incorporation. Later in the season, major differences in soil nitrates among management systems, except a peak in CT-CC-PM after 75 days of CC incorporation, were not detected until early September with all tilled systems higher than no-till. Contrary to 2016, a very low mineralization occurred in the system of plastic mulch without the green manure clover.

Soil mechanical strength is an important soil parameter that defines the level of soil compaction. As soil bulk density increases and total porosity decreases, soil resistance to root penetration increases, restricting root growth as well as water and air movement throughout the profile [26]. In our case, penetrometer readings measuring the soil strength at the end of the growing season (September) showed differences among both no-till dead mulch systems (NT-CC and NT-CC-SW) and all tilled systems in the first 5 cm of the soil profile, whereas differences in soil resistance were seen till almost 20 cm depth in 2017 (Figure 3). In both seasons, no system surpassed the 2000 kPa, the growth-limiting compaction threshold in the topsoil [27].

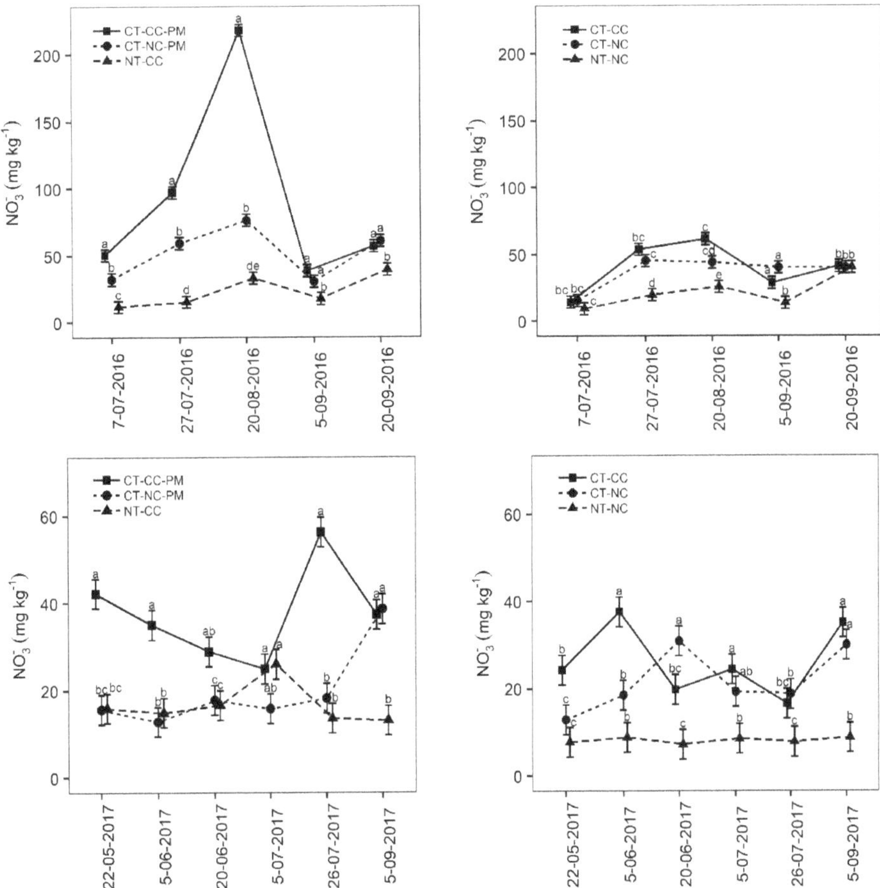

Figure 2. The soil nitrates concentration as affected by cover crop residues management in 2016 (**upper charts**) and 2017 (**lower charts**) trials. Letters of statistical significance correspond to treatments comparison within the same date of assessment. Values followed by different letters are significantly different at $p < 0.05$.

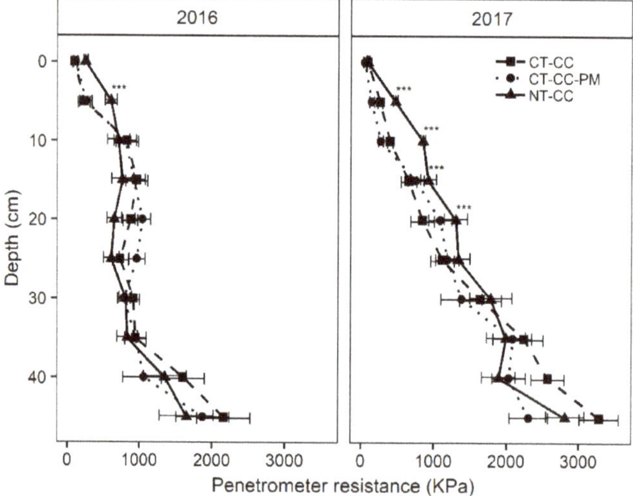

Figure 3. Penetrometer soil resistance (KPa) in the different cover crop residues management systems in 2016 (**left**) and 2017 (**right**). *** Represents statistically significant differences ($p < 0.001$).

4. Discussion

The outperformance of squarrose clover as green manure with respect to dead mulch was evident in both years of experimentation, although the positive effect of using a cover crop over a bare soil was year dependent. In fact, plant biomass, N uptake, and yield were improved where clover was incorporated as green manure in the first year of trials, but this was not noticed in the second year. This could be attributed to a lower mineralization rate in that year compared to 2016 as shown with the soil nitrates results, and their asynchrony with plant needs. Despite the higher N supply to the soil by leguminous cover crops and their capacity to improve N recovery of tomato, they can be no more effective than other cover crop species or chemical fertilizers in retaining nitrates in the soil profile, mainly due to the high mobility of nitrate ions [28–30]. Plastic mulching increases N mineralization and accumulation in soil and was reported in a large number of studies to increase crop yields, and this was mainly due to the increase in soil temperature, by 2 to 6 °C, and soil moisture as we confirmed [31–33]. In the first year of the experiment, transplantation occurred in late June which may have inflicted a thermal stress on tomato seedlings during the early growth of the plant, thus hindering their performance.

The response of organic vegetables to no-tillage conditions has not been consistent in the literature and the success seems to depend on an adequate context-specific management. Some studies showed tomato growth and production unaffected by tillage and cover crop residues management [34–36]. Other results from reduced tillage in bell pepper, onion, and zucchini production have ranged from statistically equal or even higher [37–39] to 20% and more than 90% reduction of no-till yields in these and other horticultural crops [40–42]. In our case, this could be attributed to both low soil nitrates availability and high weed competition during tomato growth. Both factors have been responsible for yields' decline in organic reduced tillage systems compared with ploughed systems in many previous experiments [43]. The slow mineralization of cover crop laid as dead mulch explained the low soil nitrates available for plants with respect to other residues management affecting plant nutrition [12–14] and partly the depression in plant performance. Nevertheless, the low mineralization may increase the N use efficiency of vegetables as demonstrated with tomato and eggplants cultivated on legume dead mulches ranging from 39 to 60% when compared to conventionally tilled systems [35,44]. Placing cover crop residues on soil surface may enhance the synchronization between N mineralized and eggplant N demand in legume cover crops, while in others (i.e., cereals) it appears to mitigate the shortage of soil inorganic N for the following vegetable [45].

One of the most important attributes of an effective mulch is biomass production with high quantity of the residues necessary for the control of an increased weed pressure, although the limit depends on the specific characteristics of the growing system [2,10,46]. Squarrose clover in our study did not exceed the 3.5 t ha^{-1} with which low performance was affected by sowing and killing date along with fluctuations in weather conditions, i.e., lower precipitations in 2017 during cover crop growth (Figure 1). The dead mulch did not ensure weed control in the first season with originally high field weed infestation, and in the next season succeeded to reduce weed infestation (38% lower weed biomass in NT-CC systems compared to NT-NC) but was not enough to increase plant performance and to decrease the competition over soil nitrates. In systems that received additional weed mowing over the dead mulch, an increase in plant performance was noticed although in some instances it was not statistically significant. Mowing, however, is not an effective measure to control weeds over the dead mulch and it disturbs the mulch and its uniformity. For these reasons, multi-tactic weed management should be considered in organic no-till as it is difficult in some cases for cover crops to be the unique method for weed control. Mechanical weed control practices that can perform on high residue conditions, a complex crop rotation and the use of allelopathic cover crops or mixtures of cover crops, are tools to be exploited in order to reduce weed pressure. The feasibility of no-till depends on field conditions and, for this, the preparation of suitable conditions before the implementation can be crucial for its success. In case of high weed seed bank, for example, stale seedbed in coordination with some previously cited practices (to mitigate the effects of frequent tillage), can be performed if possible before shifting to no-till.

Although some studies showed reduced tillage associated with a risky increase in soil compaction [47,48], our trials showed a modest soil compaction on topsoil that could not have been attributed to stress or yield depression if considering the threshold of 2000–2500 KPa for root proliferation and plant growth inhibition [27].

Tillage and cover crop residues management did not show pronounced effects on fruit basic quality where higher TSS, lower pH, and firmer fruits are preferred. This result is in accordance with other studies that showed these characteristics unaffected by tillage systems in tomato production [49,50]. However, an increase in the vitamin C content was obtained in the dead mulch system left without weed control. In previous studies, a high N concentration in the nutrient solution/fertilization was shown to favor plant leaf area development and to decrease light penetration into the canopy and the vitamin C content in fruits, what may have been found with plants from CT systems [51].

5. Conclusions

The successful implementation of conservation tillage in organic vegetable production depends on the local conditions and an adequate management to surpass the obstacles that may arise, i.e., weed pressure and soil N shortage. It may, therefore, be difficult to implement it where there is an initial high weed infestation or where a pronounced spatial variability in soil properties exist that may hinder the growth of the cover crop. Future focus should be on the design of systems that takes into account the choice of resilient productive and allelopathic cover crops, selection of suitable tomato cultivars that may withstand biotic and abiotic stresses, transplantation design (decreasing the distance between rows if possible, double rows) for a better competition with weeds, crop rotations, as well as farm machinery able to perform under no-till conditions to reduce weed pressure whenever it is necessary. Fertilization strategies targeted to supply nitrogen and other nutrients soon after transplantation of field vegetables in no-till soils should also be designed to overcome nutrient shortage due to the reduced mineralization rate and to give advantage to plants over weeds, i.e., via sub fertigation and/or mycorrhizal inoculation.

Supplementary Materials: The following are available online at http://www.mdpi.com/2073-4395/9/9/504/s1, Table S1: *p*-values for each of the factors (terms) in the variables measured.

Author Contributions: Conceptualization, D.A., C.F., A.P., and M.M.; methodology, D.A.; validation, D.A., M.M., and A.P.; formal analysis, L.M. and L.A.C.; investigation, L.A.C., M.S., D.A., and C.F.; data curation, L.M. and L.A.C.; writing—original draft preparation, L.A.C.; writing—review and editing, D.A., L.M., C.F., M.S., M.M., and A.P.; visualization, L.A.C.; supervision, D.A., M.M. and A.P.

Funding: This research received no external funding.

Acknowledgments: We would like to thank the staff of the Centro di Ricerche Agro-ambientali "Enrico Avanzi" (CiRAA) of the University of Pisa for their help in field and laboratory work. We would also like to thank Giancarlo Colelli and Maria Lucia De Chiara from the University of Foggia for the analysis of tomato fruits performed in the laboratory of the Postharvest Research Unit in the second trial.

Conflicts of Interest: The authors declare no conflict of interest.

References

1. Lernoud, J.; Willer, H. The World of Organic Agriculture. Statistics and Emerging Trends 2019. Research Institute of Organic Agriculture (FiBL), Frick, and IFOAM-Organics International, Bonn, 2019. Available online: https://shop.fibl.org/CHen/mwdownloads/download/link/id/1202/?ref=1 (accessed on 15 June 2019).
2. Peigné, J.; Ball, B.C.; Roger-Estrade, J.; David, C. Is conservation tillage suitable for organic farming? A review. *Soil Use Manag.* **2007**, *23*, 129–144. [CrossRef]
3. Lal, R.; Reicosky, D.C.; Hanson, J.D. Evolution of the plow over 10,000 years and the rationale for no-till farming. *Soil Till. Res.* **2007**, *93*, 1–12. [CrossRef]
4. Mazzoncini, M.; Antichi, D.; Di Bene, C.; Risaliti, R.; Petri, M.; Bonari, E. Soil carbon and nitrogen changes after 28 years of no-tillage management under Mediterranean conditions. *Eur. J. Agron.* **2016**, *77*, 156–165. [CrossRef]
5. Peigné, J.; Cannavaciuolo, M.; Gautronneau, Y.; Aveline, A.; Giteau, J.L.; Cluzeau, D. Earthworm populations under different tillage systems in organic farming. *Soil Till. Res.* **2009**, *104*, 207–214. [CrossRef]
6. Wittwer, R.A.; Dorn, B.; Jossi, W.; Van Der Heijden, M.G. Cover crops support ecological intensification of arable cropping systems. *Sci. Rep.* **2017**, *7*, 41911. [CrossRef] [PubMed]
7. Thorup-Kristensen, K.; Magid, J.; Jensen, L.S. Catch crops and green manures as biological tools in nitrogen management in temperate zones. *Adv. Agron.* **2003**, *79*, 227–302.
8. Gadermaier, F.; Berner, A.; Fließbach, A.; Friedel, J.K.; Mäder, P. Impact of reduced tillage on soil organic carbon and nutrient budgets under organic farming. *Renew. Agric. Food Syst.* **2012**, *27*, 68–80. [CrossRef]
9. Jokela, D.; Nair, A. No tillage and strip tillage effects on plant performance, weed suppression, and profitability in transitional organic broccoli production. *Hortscience* **2016**, *51*, 1103–1110. [CrossRef]
10. Gruber, S.; Claupein, W. Effect of tillage intensity on weed infestation in organic farming. *Soil Till. Res.* **2009**, *105*, 104–111. [CrossRef]
11. Nichols, V.; Verhulst, N.; Cox, R.; Govaerts, B. Weed dynamics and conservation agriculture principles: A review. *Field Crop. Res.* **2015**, *183*, 56–68. [CrossRef]
12. Wells, M.S.; Reberg-Horton, S.C.; Smith, A.N.; Grossman, J.M. The reduction of plant-available nitrogen by cover crop mulches and subsequent effects on soybean performance and weed interference. *Agron. J.* **2013**, *105*, 539–545. [CrossRef]
13. Soane, B.D.; Ball, B.C.; Arvidsson, J.; Basch, G.; Moreno, F.; Roger-Estrade, J. No-till in northern, western and south-western Europe: A review of problems and opportunities for crop production and the environment. *Soil Till. Res.* **2012**, *118*, 66–87. [CrossRef]
14. Berner, A.; Hildermann, I.; Fließbach, A.; Pfiffner, L.; Niggli, U.; Mäder, P. Crop yield and soil fertility response to reduced tillage under organic management. *Soil Till. Res.* **2008**, *101*, 89–96. [CrossRef]
15. WPTC. The World Processing Tomato Council. Available online: https://www.wptc.to/pdf/releases/WPTC%20world%20production%20estimate%20as%20of%2012%20February%202019.pdf (accessed on 10 August 2019).
16. Istat. Istituto Nazionale di Statistica. Available online: http://agri.istat.it/jsp/dawinci.jsp?q=plCPO0000010000013000&an=2018&ig=1&ct=418&id=15A\T1\textbar{}18A\T1\textbar{}28A (accessed on 10 August 2019).
17. Frasconi, C.; Martelloni, L.; Antichi, D.; Raffaelli, M.; Fontanelli, M.; Peruzzi, A.; Benincasa, P.; Tosti, G. Combining roller crimpers and flaming for the termination of cover crops in herbicide-free no-till cropping systems. *Plos ONE* **2019**, *14*, e0211573. [CrossRef] [PubMed]

18. Frasconi, C.; Martelloni, L.; Raffaelli, M.; Fontanelli, M.; Abou Chehade, L.; Peruzzi, A.; Antichi, D. A field vegetable transplanter for the use in both tilled and no-till soils. *Trans. ASABE* **2019**, *62*, 593–602. [CrossRef]

19. Bremner, J.M.; Mulvaney, C.S. Nitrogen-Total. In *Methods of Soil Analysis. Part 2: Chemical and Microbiological Properties*, 2nd ed.; Page, A.L., Miller, R.H., Keeney, D.R., Eds.; American Society of Agronomy, Soil Science Society of America: Madison, WI, USA, 1982; pp. 595–624.

20. Eaton, A.D.; Clesceri, L.S.; Greenberg, A.E. Determination of Anions by Ion Chromatography, Part 4000 Inorganic Nonmetallic Constituents. In *Standard Methods for the Examination of Water and Wastewater*; American Public Health Association: Washington, DC, USA, 1995.

21. Zapata, S.; DuFour, J.P. Ascorbic, dehydroascorbic and isoascorbic acid simultaneous determinations by reverse phase ion interaction HPLC. *J. Food Sci.* **1992**, *57*, 506–511. [CrossRef]

22. Gil, M.I.; Ferreres, F.; Tomás-Barberán, F.A. Effect of postharvest storage and processing on the antioxidant constituents (flavonoids and vitamin C) of fresh-cut spinach. *J. Agric. Food Chem.* **1999**, *47*, 2213–2222. [CrossRef]

23. Singleton, V.L.; Rossi, J.A. Colorimetry of total phenolics with phosphomolybdic-phosphotungstic acid reagents. *Am. J. Enol. Vitic.* **1965**, *16*, 144–158.

24. Brand-Williams, W.; Cuvelier, M.E.; Berset, C. Use of a free radical method to evaluate antioxidant activity. *LWT Food Sci. Technol.* **1995**, *28*, 25–30. [CrossRef]

25. Held, M.T.; Anthon, G.E.; Barrett, D.M. The effects of bruising and temperature on enzyme activity and textural qualities of tomato juice. *J. Sci. Food Agric.* **2015**, *95*, 1598–1604. [CrossRef]

26. Chen, G.; Weil, R.R.; Hill, R.L. Effects of compaction and cover crops on soil least limiting water range and air permeability. *Soil Till. Res.* **2014**, *136*, 61–69. [CrossRef]

27. Hamza, M.A.; Anderson, W.K. Soil compaction in cropping systems. A review of the nature, causes and possible solutions. *Soil Till. Res* **2005**, *82*, 121–145. [CrossRef]

28. Lenzi, A.; Antichi, D.; Bigongiali, F.; Mazzoncini, M.; Migliorini, P.; Tesi, R. Effect of different cover crops on organic tomato production. *Renew. Agric. Food Syst.* **2009**, *24*, 92–101. [CrossRef]

29. Dufault, R.J.; Decoteau, D.R.; Garrett, J.T.; Batal, K.D.; Granberry, D.; Davis, J.M.; Hoyt, G.; Sanders, D. Influence of cover crops and inorganic nitrogen fertilization on tomato and snap bean production and soil nitrate distribution. *J. Veg. Crop Prod.* **2000**, *6*, 13–25. [CrossRef]

30. Sainju, U.M.; Singh, B.P.; Rahman, S.; Reddy, V.R. Soil nitrate-nitrogen under tomato following tillage, cover cropping, and nitrogen fertilization. *J. Environ. Qual.* **1999**, *28*, 1837–1844. [CrossRef]

31. Teasdale, J.R.; Abdul-Baki, A.A. Soil temperature and tomato growth associated with black polyethylene and hairy vetch mulches. *J. Am. Soc. Hortic. Sci.* **1995**, *120*, 848–853. [CrossRef]

32. Ghosh, P.K.; Dayal, D.; Bandyopadhyay, K.K.; Mohanty, M. Evaluation of straw and polythene mulch for enhancing productivity of irrigated summer groundnut. *Field Crop. Res.* **2006**, *99*, 76–86. [CrossRef]

33. Hai, L.; Li, X.G.; Liu, X.E.; Jiang, X.J.; Guo, R.Y.; Jing, G.B.; Rengel, Z.; Li, F.M. Plastic mulch stimulates nitrogen mineralization in urea-amended soils in a semiarid environment. *Agron. J.* **2015**, *107*, 921–930. [CrossRef]

34. Delate, K.; Cwach, D.; Chase, C. Organic no-tillage system effects on soybean, corn and irrigated tomato production and economic performance in Iowa, USA. *Renew. Agric. Food Syst.* **2012**, *27*, 49–59. [CrossRef]

35. Campiglia, E.; Mancinelli, R.; Radicetti, E. Influence of no-tillage and organic mulching on tomato (*Solanum Lycopersicum* L.) production and nitrogen use in the mediterranean environment of central Italy. *Sci. Hortic.* **2011**, *130*, 588–598. [CrossRef]

36. Herrero, E.V.; Mitchell, J.P.; Lanini, W.T.; Temple, S.R.; Miyao, E.M.; Morse, R.D.; Campiglia, E. Use of cover crop mulches in a no-till furrow-irrigated processing tomato production system. *Horttechnology* **2001**, *11*, 43–48. [CrossRef]

37. Delate, K.; Cambardella, C.; McKern, A. Effects of organic fertilization and cover crops on an organic pepper system. *Horttechnology* **2008**, *18*, 215–226. [CrossRef]

38. Vollmer, E.R.; Creamer, N.; Reberg-Horton, C.; Hoyt, G. Evaluating cover crop mulches for no-till organic production of onions. *Hortscience* **2010**, *45*, 61–70. [CrossRef]

39. Canali, S.; Campanelli, G.; Ciaccia, C.; Leteo, F.; Testani, E.; Montemurro, F. Conservation tillage strategy based on the roller crimper technology for weed control in Mediterranean vegetable organic cropping systems. *Eur. J. Agron.* **2013**, *50*, 11–18. [CrossRef]

40. Leavitt, M.J.; Sheaffer, C.C.; Wyse, D.L.; Allan, D.L. Rolled winter rye and hairy vetch cover crops lower weed density but reduce vegetable yields in no-tillage organic production. *Hortscience* **2011**, *46*, 387–395. [CrossRef]

41. Boydston, R.A.; Williams, M.M. No-till snap bean performance and weed response following rye and vetch cover crops. *Renew. Agric. Food Syst.* **2017**, *32*, 463–473. [CrossRef]

42. Tittarelli, F.; Campanelli, G.; Leteo, F.; Farina, R.; Napoli, R.; Ciaccia, C.; Canali, S.; Testani, E. Mulch Based No-Tillage and Compost Effects on Nitrogen Fertility in Organic Melon. *Agron. J.* **2018**, *110*, 1482–1491. [CrossRef]

43. Cooper, J.; Baranski, M.; Stewart, G.; Nobel-de Lange, M.; Bàrberi, P.; Fließbach, A.; Peigné, J.; Berner, A.; Brock, C.; Casagrande, M.; et al. Shallow non-inversion tillage in organic farming maintains crop yields and increases soil C stocks: A meta-analysis. *Agron. Sustain. Dev.* **2016**, *36*, 22. [CrossRef]

44. Radicetti, E.; Mancinelli, R.; Moscetti, R.; Campiglia, E. Management of winter cover crop residues under different tillage conditions affects nitrogen utilization efficiency and yield of eggplant (*Solanum melanogena* L.) in Mediterranean environment. *Soil Tillage Res.* **2016**, *155*, 329–338. [CrossRef]

45. Radicetti, E.; Campiglia, E.; Marucci, A.; Mancinelli, R. How winter cover crops and tillage intensities affect nitrogen availability in eggplant. *Nutr. Cycl. Agroecosyst.* **2017**, *108*, 177–194. [CrossRef]

46. Teasdale, J.R.; Mohler, C.L. The quantitative relationship between weed emergence and the physical properties of mulches. *Weed Sci.* **2000**, *48*, 385–392. [CrossRef]

47. Mosaddeghi, M.R.; Mahboubi, A.A.; Safadoust, A. Short-term effects of tillage and manure on some soil physical properties and maize root growth in a sandy loam soil in western Iran. *Soil Tillage Res.* **2009**, *104*, 173–179. [CrossRef]

48. Bulan, M.T.S.; Stoltenberg, D.E.; Posner, J.L. Buckwheat species as summer cover crops for weed suppression in no-tillage vegetable cropping systems. *Weed Sci.* **2015**, *63*, 690–702. [CrossRef]

49. Thomas, R.; O'Sullivan, J.; Hamill, A.; Swanton, C.J. Conservation tillage systems for processing tomato production. *Hortscience* **2001**, *36*, 1264–1268. [CrossRef]

50. Shrestha, A.; Mitchell, J.P.; Lanini, W.T. Subsurface drip irrigation as a weed management tool for conventional and conservation tillage tomato (*Lycopersicon esculentum* Mill.) production in semi-arid agroecosystems. *J. Sustain. Agric.* **2007**, *31*, 91–112. [CrossRef]

51. Dumas, Y.; Dadomo, M.; Di Lucca, G.; Grolier, P. Effects of environmental factors and agricultural techniques on antioxidant content of tomatoes. *J. Sci. Food Agric.* **2003**, *83*, 369–382. [CrossRef]

Article

Leguminous Alley Cropping Improves the Production, Nutrition, and Yield of Forage Sorghum

Robson da Costa Leite [1,*], José Geraldo Donizetti dos Santos [2], Rubson da Costa Leite [1], Luciano Fernandes Sousa [2], Guilherme Octávio de Sousa Soares [3], Luan Fernandes Rodrigues [2], Jefferson Santana da Silva Carneiro [4] and Antonio Clementino dos Santos [2]

[1] Institute of Agricultural Sciences, Federal Rural University of Amazon (ICA-UFRA), Belém 66077-530, Pará, Brazil; rubsonif@gmail.com

[2] School of Veterinary Medicine and Animal Science, Federal University of Tocantins, Araguaina 77804-970, Tocantins, Brazil; jgsanttos@gmail.com (J.G.D.d.S.); luciano.sousa@mail.uft.edu.br (L.F.S.); luanfr@gmail.com (L.F.R.); clementino@mail.uft.edu.br (A.C.d.S.)

[3] Federal Institute of Education, Science, and Technology of Tocantins, Araguatins 77950-000, Tocantins, Brazil; guilhermeoctavio21@hotmail.com

[4] Soil Science Department, Federal University of Lavras 37200-000, Minas Gerais, Brazil; carneirojss@yahoo.com.br

* Correspondence: robsontec.agrop@gmail.com; Tel.: +55-(63)-99988-4434

Received: 24 July 2019; Accepted: 9 October 2019; Published: 14 October 2019

Abstract: This study aimed to evaluate the growth, production, and leaf contents of macronutrients, as well as the yield of forage sorghum cultivated on the alleys of *Gliricidia* (*Gliricidia sepium* (Jacq.) Kunth ex Walp.) and *Leucaena* (*Leucaena leucocephala* (Lam.) de Wit) in the presence and absence of mineral fertilization. The experiment was conducted in two different periods: During the 2016/2017 double crop (cultivation carried out at the end of the crop cycle) and during the 2017/2018 crop (cultivation carried out at the beginning of the crop cycle). A randomized block design, in which the first factor refers to cultivation systems (single sorghum, sorghum cultivated in *Gliricidia* alleys, and sorghum cultivated in *Leucaena* alleys) and the second factor refers to mineral fertilization (presence and absence of fertilization), in a 3 × 2 factorial arrangement was used. The leguminous plants were cut, and the residues were deposited in the alleys. The cultivation in alleys without mineral fertilization increased total forage biomass when compared to the single crop cultivation. Cultivation in *Leucaena* alleys showed a higher leaf content of nitrogen (N) when compared to the single crop, both in the presence and absence of mineral fertilization. In the double crop, sorghum cultivated in *Leucaena* alleys without fertilization presented a higher forage yield (up to 67%) when compared to the single crop system. However, there was no difference in yield when mineral fertilization was applied to the treatments. Overall, the alley crops were able to increase the morphological (plant height (PH), stem diameter (SD), panicle diameter (PD), and panicle length (PL)) and yield (leaf dry mass (LDM), stem dry mass (SDM), total green mass (TGM), and total dry mass TDM) variables of the crop, improving the productivity of forage sorghum.

Keywords: cultivation systems; *Gliricidia sepium*; leguminous plants; *Leucaena leucocephala*; mineral fertilization

1. Introduction

Overall, about 70% of Brazilian soils are represented by Oxisols, Ultisols and Entisols, which are soil classes of predominantly low fertility. Thus, agricultural production might be restricted if there is no nutrient addition to the soil [1]. Mineral fertilizers are often the first choice used to improve the chemical properties of soil [2,3]. However, organic materials such as plant residues can also play an important role in the improvement of tropical agriculture systems. After decomposition, the organic materials provide nutrients and substrate for the synthesis of organic matter in the soil [4]. The chemical, physical, and biological properties of soils can be greatly improved using alley cropping, which represents an accessible option for the addition of organic matter to the soil [2].

Alley cropping involves the cultivation of annual crops among the hedgerows of multipurpose trees. Plant residues from the leguminous trees can be used as organic fertilizers, promoting improvements in soil fertility [5,6]. The benefits of this system of production include surface cover with plant residues, nutrient recycling, the biological fixation of atmospheric nitrogen (BNF), and the increase of the bearing capacity of the soil [7–9].

Alley cropping is a viable option to increase biomass production per unit area. Since the plant residues can be incorporated into the soil, the transference of nutrients from trees to annual crops can also occur [8]. Furthermore, since the leguminous crops used in alleys present a deep root system, the interception of percolated nutrients along the soil profile can occur, and nutrients accumulated in layers below the root zone of annual crops can be accessed. These nutrients absorbed by the root system of the trees become inputs when transferred to the soil surface in the form of litter and other plant residues [10].

Leguminous alleys disposed in annual crops represent relevant N inputs by biological fixation, reducing the need for N fertilization. For example, *Leucaena* (*Leucaena leucocephala* (Lam.) de Wit.) and acacia (*Acacia Mangium montanum* Rumph.) arranged in maize (*Zea mays* (L.) alleys produced large amounts of N due to the increase of biomass and soil fixation [11]. There is evidence that maize cultivated in *Gliricidia* (*Gliricidia sepium* (Jacq.) Kunth ex Walp.) alleys increase their foliar N content by up to 5 g kg^{-1} when compared to single maize cultivation with mineral fertilization [12]. Legumes produce organic matter of greater bioavailability, which can also increase the cation exchange capacity (CEC) of sandy soils [13].

Among the trees and shrubs used in alley cropping systems, *Gliricidia* is widely used in the Brazilian northeast [5,8,14]. *Leucaena* is also common in alley cultivation. Though *Leucaena* has a higher competitive effect when compared to *Gliricidia*, it produces higher amounts of residues [11–15]. These species are widely used both in the incorporation of biomass into the soil and in animal feeding, and they are usually cut two to three times per year [5]. Furthermore, they are considered drought-resistant species that produce large amounts of biomass with high N levels and fast decomposition rates [16]. However, only a few scientific studies have thus far focused on the cultivation of forage sorghum (*Sorghum bicolor* (L.) Moench) in leguminous alleys, especially in areas of livestock activity [17,18].

Forage sorghum belongs to the Poaceae family, and it is among the most cultivated species in the world. Sorghum is widely used by farmers for forage production due to its high percentage of leaf and stem production when compared to other plant species. There are two categories of sorghum: Specific cultivars for grain production and specific cultivars for forage production [18]. Therefore, its high drought adaptability, high dry mass yield, and high nutrient recycling capacity make this crop attractive for forage production [13].

Agricultural crops, especially annual crops, require adequate fertility levels for their development. Therefore, the adoption of leguminous alleys in forage sorghum cultivation for sustainable soil fertility management represents a great option for nutrient input, especially for resource-poor farmers [2].

The present study was based on the hypothesis that the presence of leguminous alleys in forage sorghum cultivation would promote greater growth and development, as well as higher foliar macronutrient contents and productivity, thus making alley cropping superior to the cultivation of single sorghum. Therefore, the objective of this study was to evaluate forage sorghum cultivation using a combination of leguminous alleys and mineral fertilization.

2. Materials and Methods

2.1. Experimental Area and Treatments

Two field experiments were conducted at the School of Veterinary and Animal Science of the Federal University of Tocantins (810751.01; 9213652.69 UTM, with an altitude of 243 m), Brazil. The first experiment was implemented in the agricultural year 2016/2017 and the second in 2017/2018. This region is classified as warm and humid (AW type according to the Köppen classification). The area presents two growing seasons: A dry period with a water deficit from May to September and a rainy period between October and April [19]. The rainfall and the average temperature throughout the experimental period are shown in Figure 1.

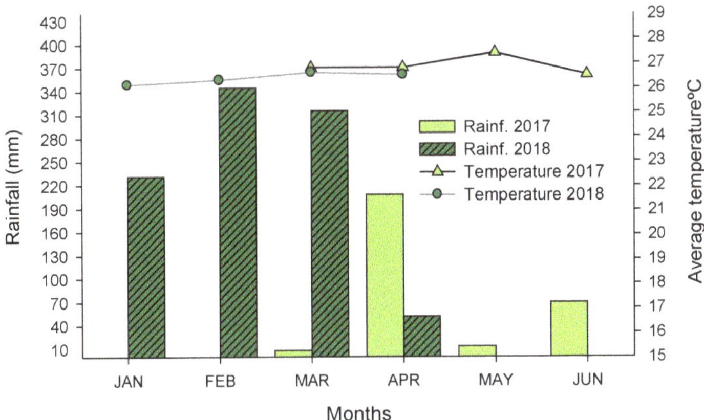

Figure 1. Average monthly rainfall and temperature for the experimental site during the 2016–2018 growing season.

Table 1 shows the physical and chemical attributes of the soil prior to the cultivation of the first crop cycle. The soil is classified as Entisol (quartzipsamment) [1].

Table 1. Data from chemical and physical analysis of the soil in preplant in the experimental site (0.00–0.20 m layer).

Attribute	
pH (H_2O)	5.3
Organic matter (g kg^{-1})	6.0
Available P (mg dm^{-3})	7.48
Available K (mg dm^{-3})	8.0
Ca^{2+} (cmol$_c$ dm^{-3})	2.47
Mg^{2+} (cmol$_c$ dm^{-3})	1.19
Al^{3+} (cmol$_c$ dm^{-3})	0.04
$H^+ + Al^{3+}$ (cmol$_c$ dm^{-3})	1.78
CEC (cmol$_c$ dm^{-3})	5.46
Sand (g kg^{-1})	893.5
Silt (g kg^{-1})	6.5
Clay (g kg^{-1})	100.0

pH (H_2O) at a ratio of 1:2.5 *m/v*; organic matter determined by the Walkley–Black method; available P e K: Mehlich-1 extraction; exchangeable Ca, Mg and Al: Extraction with KCl; H + Al: Extraction with calcium acetate; clay content: The pipette method.

The experiment followed a randomized block design with a factorial arrangement of 3 × 2 and five replications. The first factor refers to the cultivation system (single sorghum, sorghum cultivated in *Gliricidia* alleys, and sorghum cultivated in *Leucaena* alleys), and the second factor refers to mineral fertilization (the presence and absence of fertilization) (Figure 2). The total area of the experiment was 900 m^2, and each experimental unit had a total area of 30 m^2 (6 × 5 m).

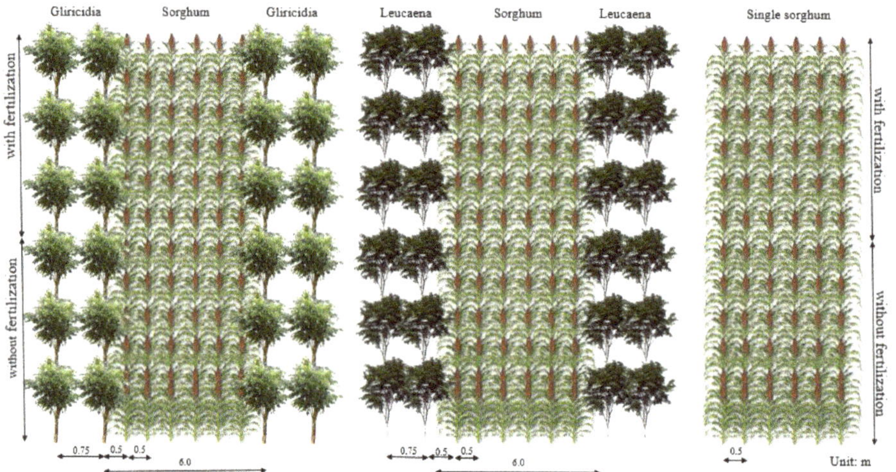

Figure 2. Scheme of the arrangement of cultivation systems and mineral fertilization in the experiment.

2.2. Establishment of Gliricidia and Leucaena Alleys and Forage Sorghum

Gliricidia and *Leucaena* were sown in 2013, with a spacing of 6 m between double rows and 0.75 m between single rows and plants. The legumes were only pruned prior to the first sorghum cultivation in 2017, and the biomass that was deposited on the soil surface was composed of leaves and stems. All plots of the same treatment received the same amount of biomass, and all available dry plant residues were added to the soil, which was added according to the dry mass content shown in Table 2. The single sorghum treatment received no plant residue.

Table 2. Macronutrient content and dry mass of plant residues deposited between lines of sorghum cultivation (double crop) in March 2017.

Legume	N (g kg^{-1})	P (g kg^{-1})	K (g kg^{-1})	Ca (g kg^{-1})	Mg (g kg^{-1})	Dry Mass (Mg ha^{-1})
Gliricidia	32.8	2.8	17.6	14.6	5.5	5.4
Leucaena	33.1	1.7	11.4	10.5	3.2	6.0

The legume biomass that was deposited in the crops was quantified using a metal frame of 0.25 m^2 and dried in a forced circulation oven at 55 °C until constant weight for chemical analysis (Table 2). One month after the cutting of the alleys, the planting furrows were manually opened. Sorghum was sown on 31 March 2017, as it was characterized as double crop. Thirteen seeds per linear meter were sown, with a spacing of 0.5 m between rows.

After collecting data on growth, yield, leaf macronutrient levels and forage sorghum productivity of the first experiment, the area remained fallow. In 2018, the alley treatments containing *Gliricidia* and *Leucaena* were again pruned, and the residues were deposited between the rows of the subsequent single sorghum cultivation, which was characterized as a crop. The experimental procedures, the cultivar, and the amount of biomass deposited were the same as the previous year, and planting was carried out on 13 January 2018.

Planting and fertilization were only carried out in plots containing mineral fertilization with nitrogen-phosphorus-potassium (NPK), according to the requirements of the crop: 20 kg ha^{-1} of N,

90 kg ha^{-1} of P$_2$O$_5$, 75 kg ha^{-1} of K$_2$O, and 30 kg ha^{-1} of micronutrients based on fritted trace elements (FTE) [20]. Fertilization was divided into two applications: When sorghum plants had four and seven fully expanded leaves by adding 100 kg ha^{-1} of N and 75 kg ha^{-1} of K$_2$O, respectively.

2.3. Analysis of Plant Tissue and Sorghum Production

When sorghum plants reached up to 50% of flowering, leaves were sampled—the fourth leaf was collected from the apex of the plants from the central rows. Eight leaves were sampled per experimental plot, and these were oven dried at 55 °C and milled in a Willey-type stationary mill for the determination of the foliar contents of N, P, K, Ca, and Mg [21].

At 85 days after sowing (DAS), eight plants of the two central rows of each plot were evaluated. Plant height and panicle length were measured from the lap of the plant up to the last expanded leaf. The stem and panicle diameters were measured using a digital caliper.

Sorghum plants were cut near the soil surface, and the plant parts were separated into stem, leaf, panicle, root, and dead material. The roots were removed with the aid of a hoe in depth of 20 cm and then washed under running water through a 2 mm sieve. Thus, the green mass of each part of the plant, as well as the leaf/stem ratio and the productivity, were obtained. The dry mass of each component was determined after drying in a forced-air oven at 55 °C until constant weight.

2.4. Statistical Analysis

All results are expressed as averages ± level of significance. The variables related to growth, production, leaf macronutrient levels, and the productivity of forage sorghum were verified for data normality by the Shapiro–Wilk test and homoscedasticity by the Bartlett test. Data were submitted to an analysis of variance and an F test, in which the averages were compared by the Tukey test at 5%.

3. Results

There was a significant interaction between the cropping systems (C) and mineral fertilization (M) for all growth variables, as well as for the dry mass of the morphological components and total sorghum production in the double crop. However, plant height (PH), panicle length (PL), and panicle dry mass (PDM) had no significant interaction between cultivation systems and mineral fertilization (Table 3). The leaf content of P and the sorghum yield in the double crop presented no interaction between the variables (C × M). Moreover, the foliar contents of N, K, and Mg also presented no interaction (C × M) (Table 4).

Table 3. Summary of the analysis of variance of the growth components as well as the dry mass of the morphological components and total forage sorghum submitted to the combination of legume alleys and mineral fertilization.

Middle Square	Double Crop										
Variation Source	PH	SD	PD	PL	LDM	SDM	PDM	DMDM	TGM	TDM	RDM
Cultivation systems (C)	687 *	0.03 *	2.5 *	34 *	318 *	6573 *	14,027 *	131 *	389,177 *	49,341 *	2108 *
Mineral fertilization (M)	4276 *	0.24 *	6.3 *	89 *	846*	27,919 *	47,521 *	632 *	1,763,023 *	192,993 *	37,439 *
(C × M)	484 *	0.008 *	0.8 *	12 *	33 *	1445 *	2102	126 *	124,341 *	9758 *	966 *
CV (%)	8.5	6.7	11.8	6.9	12.3	19.7	27.8	10.6	17.4	17.7	29.0
Crop											
Cultivation systems (C)	68.5 ns	0.2 *	0.7 *	1.1 ns	1443 *	50,903 *	658ns	155 *	777,522 *	81,883 *	8905 *
Mineral fertilization (M)	16450 *	1.3 *	12.4 *	150 *	19,885 *	793,331 *	67,794 *	2353 *	1,590,2883 *	1,797,184 *	178,427 *
(C × M)	55.1 ns	0.01 *	0.2 *	1.2 ns	841 *	93 *	1028 ns	112 *	82,518 *	3378 *	8815 *
CV (%)	8.0	12.3	9.8	4.1	15.9	17.0	15.1	25.2	12.1	13.7	26.1

Plant height (PH), stem diameter (SD), panicle diameter (PD), panicle length (PL), leaf dry mass (LDM), stem dry mass (SDM), panicle dry mass (PDM), dry mass of the dead material (DMDM), total green mass (TGM), total dry mass (TDM), and root dry mass (RDM). * = Significant at 5% probability; ns = Not significant.

Table 4. Summary of the analysis of variance of the stem leaf ratio, foliar macronutrient content, and forage sorghum yield submitted to the combination of alleys and mineral fertilization.

Middle Square	Double Crop						
Variation Source	Stem/Leaf Relation	N	P	K	Ca	Mg	Yield
Cultivation systems (C)	0.1 *	1.7 *	0.01 ns	70.3 *	0.6 *	0.5 *	5259 *
Mineral fertilization (M)	0.2 *	0.004 ns	0.007 ns	37.8 *	0.001 *	0.3 *	123,413 *
(C × M)	0.06 *	0.05 *	0.001 ns	52.6	0.08 *	0.3 *	167,987 *
CV (%)	26.8	7.0	3.4	12.7	13.3	21.3	16.4
Crop							
Cultivation systems (C)	0.01 *	85.8 *	0.7 *	0.5 ns	0.5 *	0.09 *	218,755 *
Mineral fertilization (M)	0.19*	0.07 ns	0.02 ns	120 *	0.5 *	0.5 *	104,247 *
(C × M)	0.002 *	1.4 ns	0.05 *	2.8 ns	0.1 *	0.006 ns	443,964 *
CV (%)	14.3	10.8	14.3	9.2	13.8	15.0	12.2

* Significant at 5% probability; ns = Not significant.

3.1. Sorghum Growth

Except for the panicle diameter under the effect of the *Leucaena* alleys in the double crop (in which no alteration was verified), the components related to sorghum growth (plant height, stem diameter, panicle diameter and panicle length) were higher in the presence of fertilization when compared to the non-fertilized treatments (Figure 3).

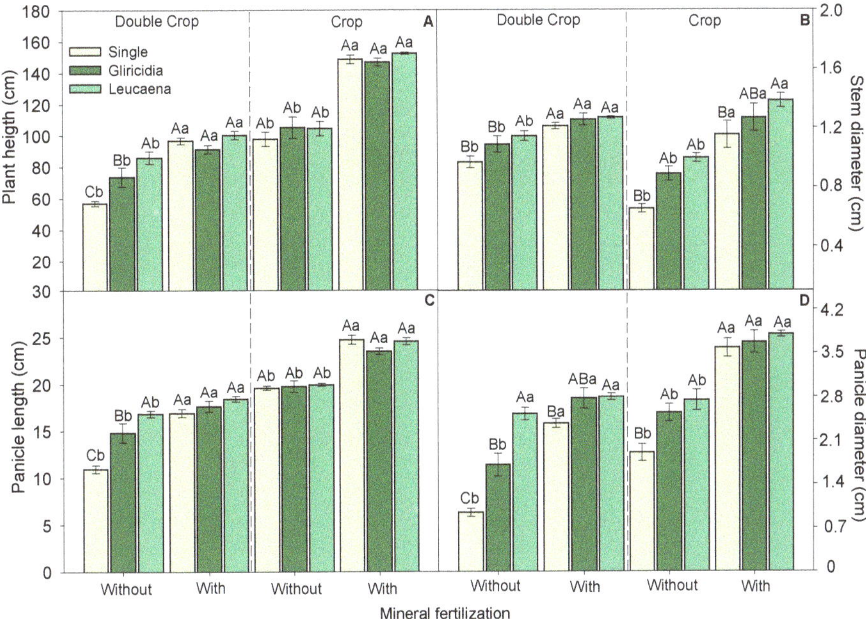

Figure 3. Plant height (**A**), stern diameter (**B**), panicle length (**C**) and panicle diameter (**D**) of forage sorghum submitted to the combination of leguminous alleys and mineral fertilization. Averages followed by the same lowercase letter for fertilization and uppercase letter for cropping systems were found by the Tukey test to not differ from each other at 5%.

In the crop, the presence of *Leucaena* alleys increased the diameters of the stem and the panicle when compared to the single sorghum system. In the double crop, *Leucaena* cultivation caused more positive impacts on plant height, stem diameter, panicle diameter, and panicle length (up to 72% more than single sorghum) than in the experimental units with the presence of mineral fertilization. In the experimental units without mineral fertilization, the cultivation in *Leucaena* alleys was 30% superior to the single sorghum cultivation plots in relation to plant height, stem diameter, panicle diameter and panicle length.

3.2. Morphological Components and Biomass Production of Forage Sorghum

In the absence of mineral fertilization, the addition of the plant residues increased the leaf dry mass (LDM) and the stem dry mass (SDM) of sorghum when compared to the single crop cultivation. This beneficial effect was verified both in the double crop and in the crop. However, there were no alterations between the cultivation systems in the presence of mineral fertilization (Figure 4). On average, leaf dry mass in alley crops exceeded the single crop by 28% and 66% in the double crop and crop, respectively. The potential of SDM production under the effect of the leguminous alleys was, on average, 100% higher than single sorghum in the double crop and crop (Figure 4). Except for sorghum in the double crop cultivated in *Leucaena* alleys, mineral fertilization increased the LDM and the SDM in relation to the treatments without fertilization.

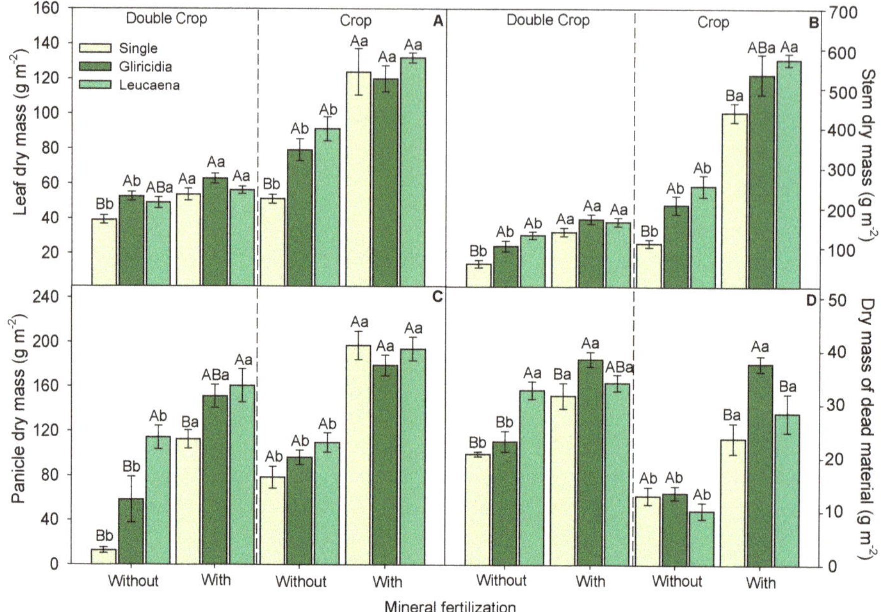

Figure 4. Leaf dry mass (**A**), stem dry mass (**B**), panicle dry mass (**C**), and dry mass of dead material (**D**) of forage sorghum submitted to the combination of leguminous alleys and mineral fertilization. Averages followed by the same lowercase letter for fertilization and uppercase letter for cropping systems were found by the Tukey test to not differ from each other at 5%.

Regarding the production of the PDM in the double crop, sorghum cultivated in *Leucaena* alleys was much higher than single sorghum (up to 500%). *Leucaena* alleys influenced the increase of the dry mass of dead material (DMDM) of sorghum cultivated without fertilization in the double crop. However, with the application of mineral fertilization, the presence of *Gliricidia* alleys caused an increase in the double crop and crop of 17% and 30%, respectively, when compared to the single crop.

As for the total green mass (TGM) produced by sorghum in the experiment conducted during the crop, *Gliricidia* and *Leucaena* alleys without fertilization and with fertilization were 78% and 11% higher than single sorghum, respectively. In the double crop in the absence of mineral fertilization, the greater green mass production of sorghum cultivated between the alleys (up to 116% when compared to single sorghum cultivation) was also verified. Nevertheless, there was no effect of the cultivation system when mineral fertilization was applied (Figure 5). In general, the total dry mass (TDM) followed the same patterns of green mass; however, in the double crop without the application of mineral fertilization, *Leucaena* alleys caused more benefits to sorghum development than *Gliricidia* alleys.

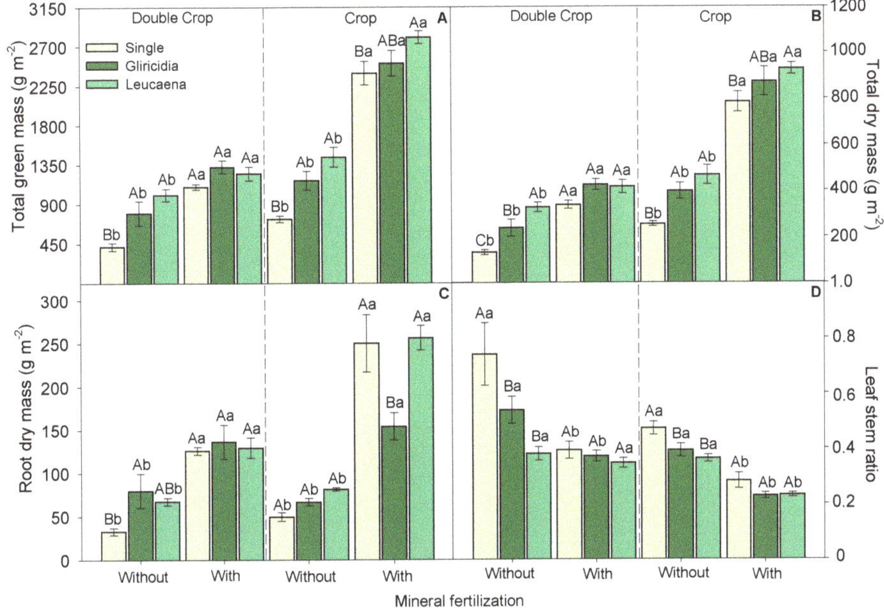

Figure 5. Total green mass (**A**), total dry mass (**B**), root dry mass (**C**) and forage sorghum leaf stem ratio (**D**) of forage sorghum submitted to the combination of leguminous alleys and mineral fertilization. Averages followed by the same lowercase letter for fertilization and uppercase letter for cropping systems were found by the Tukey test to not differ from each other at 5%.

Both in the crop and in the double crop, the root dry mass (RDM) was strongly influenced by the application of mineral fertilization with an increase in root production, regardless of the cultivation system. However, in the absence of mineral fertilization in the double crop, the sorghum RDM in *Gliricidia* alleys exceeded single sorghum by up to 142%.

The absence of mineral fertilization increased the leaf steam ratio regardless of the cultivation system. The cultivation in alleys without fertilization presented leaf stem ratios 38% and 20% lower than the single crop in the double crop and crop, respectively.

3.3. Leaf Macronutrient Contents

In both the double crop and in the crop, fertilization did not alter the contents of N and P in sorghum plants. However, in the crop, the cultivation of *Leucaena* alleys caused an increase in the contents of these nutrients when compared to the single crop (up to 28% and 26% for N and P, respectively). However, in the double crop with the presence of mineral fertilization, a lower N content was observed under the effect of *Leucaena* alleys (Table 5). A lower leaf content of P also predominated under the effect of the alleys when compared to the single crop.

As for the leaf content of K, the cultivations of the alleys without the presence of mineral fertilization in the double crop were benefited by 50% under the effect of *Gliricidia* and by 100% under the effect of *Leucaena*. However, in the crop, mineral fertilization was responsible to increase the content of K, regardless of the cultivation system.

Gliricidia and *Leucaena* alley cultivation predominantly provided the lowest leaf contents of Ca and Mg in both the double crop and the crop when compared to the single crop.

Table 5. Leaf macronutrients contents of forage sorghum cultivated with the combination of alleys and mineral fertilization.

Cultivation System	Mineral Fertilization														
	Absent	Present	Average	Absent	Present	Average	Absent	Present	Average	Absent	Present	Average	Absent	Present	Average
	N (g kg^{-1})			P (g kg^{-1})			K (g kg^{-1})			Ca (g kg^{-1})			Mg (g kg^{-1})		
Double crop															
Single	7.2 Aa	7.3 Aa	-	1.6 Aa	1.6 Aa	1.67A	10.8 Cb	17.9 Aa	-	1.46 Aa	1.26 Ab	-	1.2 Aa	0.6 Ab	-
Gliricidia	6.8 Aa	6.7 ABa	-	1.5 Aa	1.6 Aa	1.59B	16.2 Ba	18.0 Aa	-	0.95 Ba	1.1 ABa	-	0.7 Ba	0.6 Aa	-
Leucaena	6.5 Aa	6.3 Ba	-	1.5 Aa	1.6 Aa	1.61B	20.7 Aa	18.6 Aa	-	0.81 Ba	0.9 Ba	-	0.4 Ca	0.5 Aa	-
Average	-	-	-	1.6 a	1.6 a	-	-	-	-	-	-	-	-	-	-
Crop															
Single	20.9 Ba	20.0 Ba	20.5 B	2.1 Aa	1.8 Ba	-	18.3 Ab	21.1 Aa	19.7 A	1.4 Aa	1.1 Ab	-	0.9 Aa	0.6 Ab	0.78 A
Gliricidia	23.8 ABa	24.1 ABa	23.9 A	2.0 Aa	2.0 Ba	-	16.8 Ab	21.7 Aa	19.2 A	1.1 Ba	0.7 Bb	-	0.7 Ba	0.4 Bb	0.5 B
Leucaena	26.1 Aa	26.5 Aa	26.3 A	2.4 Aa	2.49 Aa	-	17.3 Ab	21.6 Aa	19.4 A	0.8 Ca	0.7 Ba	-	0.8 ABa	0.5 ABb	0.7 AB
Average	23.6 a	23.5 a	-	-	-	-	17.4 b	21.4 a	-	-	-	-	0.8 a	0.5 b	-

Averages followed by the same lowercase letter within the line and uppercase letter within the column were found by the Tukey test to not differ from each other at 5%.

3.4. Sorghum Yield

The forage sorghum yield was improved in the crop by the mineral fertilization, regardless of the cultivation system. However, single sorghum cultivation was more dependent on fertilization than sorghum cultivated in the presence of the alleys. Single sorghum without fertilization had its productivity decreased by 62% when compared to the cropping systems with mineral fertilization in the double crop, while in the presence of the alleys, the decrease was 30%. Regarding the cultivation of the crop the productivity decreased by 69% for single sorghum and by 48% for alley cropping.

In the crop without mineral fertilization, the crop systems presented similar productivities. However, with the presence of fertilization, single sorghum productivity increased by an average of 5500 kg ha^{-1} when compared to cultivation in *Gliricidia* and *Leucaena* alleys (Figure 6).

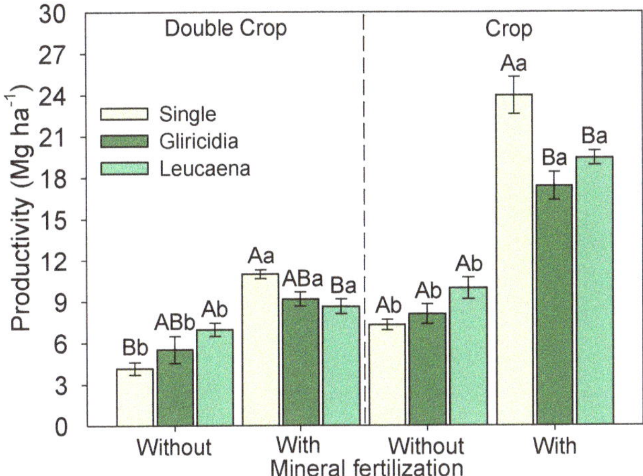

Figure 6. Productivity of forage sorghum cultivated with the combination of alleys and mineral fertilization. Averages followed by the same lowercase letter for fertilization and uppercase letter for cropping systems were found by the Tukey test to not differ from each other at 5%.

In the double crop without mineral fertilization, sorghum cultivation in *Leucaena* alleys presented a higher productivity than the single sorghum cultivation (up to a 67% increase). With the presence of mineral fertilization, sorghum productivity in *Gliricidia* alleys was similar to single sorghum. Nevertheless, the cultivation in *Leucaena* alleys was smaller than single sorghum.

4. Discussion

4.1. Morphological Components of Growth and Biomass Production of Forage Sorghum

Plant heights, stem diameters, panicle diameters, and panicle lengths are characteristics that positively influence the production of sorghum [18]. Studies on the benefits of corn and sorghum alleys have found evidence that leguminous alleys increase plant height when compared to single cultivation [22,23]. The application of leguminous residues controls weeds and improves the physical, chemical, and biological properties of the soil [24].

The dry mass of the morphological components of the alleys was increased when compared to single sorghum, especially in the absence of mineral fertilization. The distinct characteristics of the legumes and sorghum resulted in the exploration of the different layers of the soil (as well as soil structuring and dry mass production), which is associated with a lower rate of the decomposition of residues and nutrient recycling that benefits agricultural crop [25].

The total green mass and the total dry mass of the alley cropping systems were higher than the single sorghum system, except in the double crop with the presence of mineral fertilization, in which the crop systems did not cause changes in these variables. Plants of the Poaceae family grown in legume alleys increase the production of green mass and dry mass when compared to conventional cultivation without alleys [26]. The most important advantages of the alley cropping system in relation to single crops are: An increased production of green and dry mass, a greater accumulation of nutrients, and soil protection [22].

When studying maize cultivation in *Gliricidia* alleys [13], we observed a higher total dry mass of the crop in alleys when compared to maize cultivation fertilized with manure and conventional maize cultivation. Due to their high capacity to fix atmospheric N and to produce biomass under conditions of low water availability, *Gliricidia* and *Leucaena* alleys are capable of improving soil fertility and increasing the dry mass production of plants in the Poaceae family [27].

The cultivation systems without the presence of mineral fertilization were similar regarding root dry mass, whereas sorghum cultivated in *Gliricidia* alleys was superior to the single sorghum treatment in the double crop. However, with the presence of mineral fertilization in the crop, the RDM of sorghum in *Gliricidia* alleys was lower than the other cropping systems, which shows that the plant did not require as much investment in roots. The production of sorghum roots may depend on competitiveness with the legumes arranged in alleys, and the longer the establishment time of the alleys, the greater the competitiveness of the legumes with the crop [5–28].

The stem leaf ratio is important for the quality of the forage. In the absence of mineral fertilization, the systems of cultivations in leguminous alleys exerted influence on sorghum development, thus resulting in lower leaf proportion and impacting the lower leaf stem ratio [29].

4.2. Macronutrient Leaf Contents and Productivity

The *Leucaena* alleys cultivation system with and without the mineral fertilization of the crop was the only one able to provide a foliar content of N within the critical level for the production of 80% of the crop potential [21]. However, the leaf N level of the double crop was low for an adequate crop production; this was related to the scarcity of rain, which caused a limited N availability to the plants [30].

When studying *Gliricidia* and *Leucaena* alleys as a way to improve soil properties, Fernandes et al. [27] found that the residues incorporated 160 and 130 kg ha^{-1} year^{-1} of N, respectively (when only considering N). The leguminous alleys recovered about 20% of N directly from the residues deposited in the soil [11–16]. BNF can also represent N inputs relevant to the soil/plant system and reduce the need for N fertilizer application, which is often expensive and most susceptible to losses [8].

In the experiment developed during the crop of 2017/2018, the cultivation in alleys was superior when to that of the single crop for N leaf content. [30,31] found that leguminous alleys could increase the efficiency of N fertilizer use. However, in the present study, mineral fertilization did not influence the leaf N content of the crop.

In the present study, *Leucaena* alley cropping was the only system that increased the leaf concentration of N in sorghum, thus contributing to an increase of 28%. When studying maize cultivation in *Gliricidia* alleys [12], we found an increase of 86% of N of the particulate organic matter of the soil in relation to the cultivation of single maize, which was compared to the effect of the use of 50 kg ha^{-1} of N fertilizer.

As for the content of leaf P, the crop cultivation systems were adequate, except for single sorghum with mineral fertilization. However, in the double crop, even with the presence of mineral fertilization, the P levels of the cropping systems were below the ideal for the sorghum crop [19]. De Paula et al. [32] found that P from the decomposition of leguminous residues formed less water-soluble compounds and moved more slowly from one compartment to another. Furthermore, the half-life of nutrient release is shorter in the double crop.

The foliar content of P of the double crop and crop did not change with the mineral fertilization factor. [33] pointed out that BNF carried out by legumes results in higher energy and P expenditure by legumes. However, in the crop of 2017/2018, sorghum cultivated in *Leucaena* alleys with mineral fertilization showed a higher leaf content of P when compared to the other cropping systems. Nevertheless, there were no alterations between crop systems without mineral fertilization.

In a nutritional study of corn intercropped with legumes, the authors of [26] found changes in leaf P content only in the second year, in which this content was higher in legume crops than the conventional treatment. Thus, several plant species, especially perennial legumes, can use non-labile fractions of P by modifying the chemistry of their rhizosphere, excreting protons and organic acids to solubilize P and leave it available for crops [34].

The content of K remained adequate in all cropping systems in the two years of cultivation, regardless of the mineral fertilization, except for the single sorghum cultivation without mineral fertilization of the double crop, which presented a K content below suitable levels for the crop [21].

In the legume alley cropping without the mineral fertilization of the double crop, the content of leaf K was higher than that of the single crop. However, with the presence of mineral fertilization, the crop systems were similar. These results indicate that, in addition to providing nutrients from the plant residues to the main crop, these legumes probably recycled K from depths beyond the crop zone by sorghum roots [35,36].

The cropping systems did not undergo alterations regarding the foliar content of K, either with fertilization or without fertilization. Since the crop experiment was implemented in the second cycle, there is evidence that the content of K originating from the first cycle was sufficient and altered the effect in the double crop cycle [33]. The authors of [37] stated that in the cultivation of maize in alleys, legumes positively increased the content of leaf K in maize from the first crop cycle, which was similar to the conventional cultivation.

As for the levels of Ca and Mg, they were inadequate in all treatments—both in the double crop and in the crop [21]. Regardless of mineral fertilization, the lowest foliar contents of Ca and Mg were verified in the presence of leguminous alley crops. This result denotes the existence of competition between the legumes and the sorghum crop. However, the competition increases with the presence of mineral fertilization in the crop. The legume species require the same resources as the associated crops, which can result in both complementarity and competition [5].

Legumes have deep roots, can intercept percolated nutrients along the soil profile, and can access nutrients accumulated in the layers below the root zone of annual crops. These nutrients absorbed by the root system of the trees become inputs in the form of plant residues [10]. In general, legume residues provide Ca and Mg for agricultural crops. However, the slow release of these nutrients is probably due to the fact that they are some of the constituents of the middle lamella of the cell wall, forming one of the most recalcitrant components of the tissues [28–38].

On average, the presence of mineral fertilization in the cropping systems doubled the productivity in the two years of experiment. The leguminous alleys of the crop without mineral fertilization showed results similar to single sorghum cultivation. As for the presence of mineral fertilization, the systems of cultivation in leguminous alleys were smaller than the single sorghum cultivation.

Akinnifesi et al. [30] reported that corn yield in *Gliricidia* alleys without the mineral fertilization of N and P was 39% higher than in the single maize plots that received the recommended total amounts of N and P. When the *Gliricidia* alleys were altered with 50% N and 100% P, the yield increased by 79%.

Crops in legume alleys presented a surface area 30% lower than the single crop. In a study that related sorghum cultivation to leguminous alleys, the authors of [39] pointed out that the yield of sorghum cultivated in leguminous alleys corresponded to 94% in relation to conventional cultivation, although 86% of the area was occupied in the system.

The cultivation in leguminous alleys denotes its importance as a practice of agriculture with low external input as a form of soil fertilization, because it can maintain or increase the productive capacity of integrated agricultural crops [33]. Leguminous alleys are important for the morphological development, growth, biomass production, and nutrition components of forage sorghum, especially N. In the present study, sorghum yield was increased and presented a direct relation with the presence of the alleys.

5. Conclusions

The presence of mineral fertilization improved the results of all studied cultivation systems (single sorghum, sorghum grown in *Leucaena* alleys and sorghum grown in *Gliricidia* alleys) when compared to the absence of fertilization. Nevertheless, the cultivation in alleys was not significantly different when considering the influence of mineral fertilization on plant height, panicle length, leaf dry mass, leaf stem ratio and the leaf content of K in both experimental periods. *Leucaena* alleys outperformed *Gliricidia* alleys considering plant height, panicle diameter and panicle length. At least one of the alley cropping systems was larger than the single sorghum for the dry mass of the morphological components, total yield of green mass, and dry and root mass.

The contents of leaf N and P were not related to mineral fertilization. The alleys were able to positively influence the contents of these nutrients in the plant. The cultivation in *Leucaena* alleys increased the content of P when compared to single sorghum in the crop. The cultivation of sorghum in the alley cropping system also contributed to a higher leaf content of K when compared to the single crop in the double crop. The leguminous plants could have competed with sorghum for Ca and Mg, resulting in lower levels of these elements when compared to the single crop.

Mineral fertilization increased sorghum productivity regardless of the studied cultivation system. *Gliricidia* and *Leucaena* alleys showed clear potential to increase sorghum productivity, especially for sorghum cultivated in the double crop. However, in the crop with the presence of mineral fertilization, the cultivation in alleys did not overcome single sorghum cultivation regarding productivity.

Author Contributions: Conceptualization, R.d.C.L. (Robson da Costa Leite) and A.C.d.S.; methodology, R.d.C.L. (Robson da Costa Leite), R.d.C.L. (Rubson da Costa Leite) and A.C.d.S.; software, R.d.C.L. (Robson da Costa Leite), L.F.S. and L.F.R.; validation, A.C.d.S., J.G.D.d.S. and L.F.S.; data curation, R.d.C.L. (Robson da Costa Leite), G.O.d.S.S., J.S.d.S.C. and R.d.C.L. (Rubson da Costa Leite); writing—original draft preparation, R.d.C.L. (Robson da Costa Leite); writing—review and editing, R.d.C.L. (Robson da Costa Leite) and J.G.D.d.S.; supervision; A.C.d.S.

Funding: This study was carried out with the support of the Coordination of Improvement of Higher Education Personnel-Brazil (CAPES)-Financing Code 001.

References

1. EMBRAPA (Empresa Brasileira de Pesquisa Agropecuária). *Sistema Brasileiro de Classificação de Solos*, 3rd ed.; Centro Nacional de Pesquisa de Solos: Brasília, Brazil, 2013; p. 353.

2. Ekepu, D.; Tirivanhu, P. Assessing socio-economic factors influencing adoption of legume-based multiple cropping systems among smallholder sorghum farmers in Soroti, Uganda. *S. Afr. J. Agric. Ext. (SAJAE)* **2016**, *44*, 195–215. [CrossRef]

3. Omari, R.A.; Aung, H.P.; Hou, M.; Yokoyama, T.; Onwona-Agyeman, S.; Oikawa, Y.; Fujii, Y.; Bellingrath-Kimura, S.D. Influence of Different Plant Materials in Combination with Chicken Manure on Soil Carbon and Nitrogen Contents and Vegetable Yield. *Pedosphere* **2016**, *26*, 510–521. [CrossRef]

4. Williams, D.M.; Blanco-Canqui, H.; Francis, C.A.; Galusha, T.D. Organic Farming and Soil Physical Properties: An Assessment after 40 Years. *Agron. J.* **2017**, *109*, 600. [CrossRef]

5. Marin, A.M.P.; Menezes, R.S.C.; Salcedo, I.H. Produtividade de milho solteiro ou em aléias de gliricídia adubado com duas fontes orgânicas. *Pesquisa Agropecuária Brasileira* **2007**, *42*, 669–677. [CrossRef]

6. Oliveira, F.R.A.D.; Souza, H.A.D.; Carvalho, M.A.R.D.; Costa, M.C.G. Green fertilization with residues of leguminous trees for cultivating maize in degraded soil. *Rev. Caatinga* **2018**, *31*, 798–807. [CrossRef]

7. De Moura, E.G.; Albuquerque, J.M.; Aguiar, A.D.C.F. Growth and productivity of corn as affected by mulching and tillage in Alley cropping systems. *Sci. Agricola* **2008**, *65*, 204–208. [CrossRef]

8. Martins, J.C.R.; Freitas, A.D.S.D.; Menezes, R.S.C.; Sampaio, E.V.D.S.B. Nitrogen symbiotically fixed by cowpea and gliricidia in traditional and agroforestry systems under semiarid conditions. *Pesquisa Agropecuária Brasileira* **2015**, *50*, 178–184. [CrossRef]

9. De Oliveira, V.R.; E Silva, P.S.L.; Pontes, F.S.T.; De Paiva, H.N.; Antonio, R.P. Growth Of Arboreal Leguminous Plants And Maize Yield In Agroforestry Systems. *Revista Árvore* **2016**, *40*, 679–688. [CrossRef]

10. Loss, A.; Pereira, M.G.; Ferreira, E.P.; Dos Santos, L.L.; Beutler, S.J.; Júnior, A.S.D.L.F. Frações oxidáveis do carbono orgânico em argissolo vermelho-amarelo sob sistema de aleias. *Revista Brasileira de Ciência do Solo* **2009**, *33*, 867–874. [CrossRef]

11. Aguiar, A.D.C.F.; Amorim, A.P.; Coelho, K.P.; De Moura, E.G. Environmental and agricultural benefits of a management system designed for sandy loam soils of the humid tropics. *Revista Brasileira de Ciência do Solo* **2009**, *33*, 1473–1480. [CrossRef]

12. Beedy, T.; Snapp, S.; Akinnifesi, F.; Sileshi, G. Impact of Gliricidia sepium intercropping on soil organic matter fractions in a maize-based cropping system. *Agric. Ecosyst. Environ.* **2010**, *138*, 139–146. [CrossRef]

13. Primo, D.C.; Menezes, R.S.C.; Sampaio, E.V.D.S.B.; Garrido, M.D.S.; Junior, J.C.B.D.; Souza, C.S. Recovery of N applied as 15N-manure or 15N-gliricidia biomass by maize, cotton and cowpea. *Nutr. Cycl. Agroecosyst.* **2014**, *100*, 205–214. [CrossRef]

14. Primo, D.; Menezes, R.; Filho, R.; Dutra, E.; Silva, E.; Alves, R.; Sampaio, E.; Antonino, A.; Lucena, E. Characteristics Physico-Chemical and Carbon Balance in Fluvic Entisol after Six Years Fertilization with Manure and Gliricidia. *J. Exp. Agric. Int.* **2018**, *23*, 1–11. [CrossRef]

15. Moura-Silva, A.G.; Aguiar, A.D.C.F.; Jorge, N.; Agostini-Costa, T.D.S.; Moura, E.G. Food quantity and quality of cassava affected by leguminous residues and inorganic nitrogen application in a soil of low natural fertility of the humid tropics. *Bragantia* **2017**, *76*, 406–415. [CrossRef]

16. Mundus, S.; Menezes, R.S.C.; Neergaard, A.; Garrido, M.S. Maize growth and soil nitrogen availability after fertilization with cattle manure and/or gliricidia in semi-arid NE Brazil. *Nutr. Cycl. Agroecosyst.* **2008**, *82*, 61–73. [CrossRef]

17. Calvo, C.L.; Foloni, J.S.S.; Brancalião, S.R. Produtividade de fitomassa e relação C/N de monocultivos e consórcios de guandu-anão, milheto e sorgo em três épocas de corte. *Bragantia* **2010**, *69*, 77–86. [CrossRef]

18. De Oliveira, L.B.; Pires, A.J.V.; Viana, A.E.S.; Matsumoto, S.N.; De Carvalho, G.G.P.; Ribeiro, L.S.O. Produtividade, composição química e características agronômicas de diferentes forrageiras. *Revista Brasileira de Zootecnia* **2010**, *39*, 2604–2610. [CrossRef]

19. Alvares, C.A.; Stape, J.L.; Sentelhas, P.C.; Gonçalves, J.L.D.M.; Sparovek, G. Köppen's climate classification map for Brazil. *Meteorol. Z.* **2013**, *22*, 711–728. [CrossRef]

20. Sousa, D.M.G.; Lobato, E. *Cerrado: Correção do Solo E Adubação*, 2nd ed.; Embrapa Cerrados: Brasília, Brazil, 2004; p. 416.

21. Boaretto, A.E.; van Raij, B.; da Silva, F.C.; Chitolina, J.C.; Tedesco, M.J.; do Carmo, C.A.F. Amostragem acondicionamento e preparo de amostras de planta para análise química. In *Manual de Análises Químicas de Solos, planta e fertilizantes*; da Silva, F.C., Ed.; Embrapa informação tecnológica: Brasília, Brazil, 2009; pp. 61–85.

22. Andrade Neto, R.C.; Miranda, N.O.; Duda, G.P.; Góes, G.B.; Lima, A.S. Crescimento e produtividade do sorgo forrageiro BR 601 sob adubação verde. *Revista Brasileira de Engenharia Agrícola e Ambiental* **2010**, *14*, 124–130. [CrossRef]

23. Silva, A.R.D.; Collier, L.S.; Flores, R.A.; Santos, V.M.D.; Silva, L.L.D.; Oliveira, V.A.; Barbosa, J.M. Productive Yield of Cowpea and Maize in Single Crop and Mixtures in an Agroforestry System. *Am. Eur. J. Agric. Environ. Sci.* **2015**, *15*, 85–92.

24. Silva, P.S.L.; Oliveira, V.R.D.; Silva, P.I.B.; Chicas, L.D.S.; Tomaz, F.L.D.S. Effects of ground cover from branches of arboreal species on weed growth and maize yield. *Revista Ciência Agronômica* **2015**, *46*, 809–817. [CrossRef]

25. De Carvalho, W.P.; De Carvalho, G.J.; Neto, D.D.O.A.; Teixeira, L.G.V. Desempenho agronômico de plantas de cobertura usadas na proteção do solo no período de pousio. *Pesquisa Agropecuária Brasileira* **2013**, *48*, 157–166. [CrossRef]

26. Heinrichs, R.; Vitti, G.C.; Moreira, A.; Fancelli, A.L. Produção e estado nutricional do milho em cultivo intercalar com adubos verdes. *Revista Brasileira de Ciência do Solo* **2002**, *26*, 225–230. [CrossRef]

27. Fernandes, M.F.; Barreto, A.C. Cultivo de Gliricidia sepium e Leucaena leucocephala em alamedas visando a melhoria dos solos dos tabuleiros costeiros. *Pesquisa Agropecuária Brasileira* **2001**, *36*, 1287–1293.

28. Martins, J.C.R.; Menezes, R.S.C.; Sampaio, E.V.S.B.; Dos Santos, A.F.; Nagai, M.A. Produtividade de biomassa em sistemas agroflorestais e tradicionais no Cariri Paraibano. *Revista Brasileira de Engenharia Agrícola e Ambiental* **2013**, *17*, 581–587. [CrossRef]

29. Gama, T.D.C.M.; Volpe, E.; Lempp, B.; Galdeia, E.C. Recuperação de pasto de capim-braquiária com correção e adubação de solo e estabelecimento de leguminosas. *Revista Brasileira de Saúde e Produção Animal* **2013**, *14*, 635–647. [CrossRef]

30. Akinnifesi, F.K.; Makumba, W.; Sileshi, G.; Ajayi, O.C.; Mweta, D. Synergistic effect of inorganic N and P fertilizers and organic inputs from Gliricidia sepium on productivity of intercropped maize in Southern Malawi. *Plant Soil* **2007**, *294*, 203–217. [CrossRef]

31. Chirwa, P.W.; Black, C.R.; Ong, C.K.; Maghembe, J. Nitrogen Dynamics in Cropping Systems in Southern Malawi Containing Gliricidia sepium, Pigeonpea and Maize. *Agrofor. Syst.* **2006**, *67*, 93–106. [CrossRef]

32. De Paula, P.D.; Campello, E.F.C.; Guerra, J.G.M.; Santos, G.D.A.; De Resende, A.S. DECOMPOSIÇÃO DAS PODAS DAS LEGUMINOSAS ARBÓREAS Gliricidia sepium E Acacia angustissima EM UM SISTEMA AGROFLORESTAL. *Ciência Florest.* **2015**, *25*, 791–800. [CrossRef]

33. Queiroz, L.R.; Coelho, F.C.; Barroso, D.G.; Galvão, J.C.C. Cultivo de milho consorciado com leguminosas arbustivas perenes no sistema de aléias com suprimento de fósforo. *Rev. Ceres.* **2008**, *55*, 409–415.

34. Hauggaard-Nielsen, H.; Jensen, E.S. Facilitative Root Interactions in Intercrops. *Plant Soil* **2005**, *274*, 237–250. [CrossRef]

35. Makumba, W.; Janssen, B.; Oenema, O.; Akinnifesi, F.; Mweta, D.; Kwesiga, F. The long-term effects of a gliricidia–maize intercropping system in Southern Malawi, on gliricidia and maize yields, and soil properties. *Agric. Ecosyst. Environ.* **2006**, *116*, 85–92. [CrossRef]

36. Primo, D.C.; Menezes, R.S.C.; Oliveira, F.F.D.; Dubeux Júnior, J.C.B.; Sampaio, E.V.S.B. Timing and placement of cattle manure and/or gliricidia affects cotton and sunflower nutrient accumulation and biomass productivity. *Anais da Academia Brasileira de Ciências* **2018**, *90*, 415–424. [CrossRef] [PubMed]

37. De Moura, E.G.; Serpa, S.S.; Dos Santos, J.G.D.; Sobrinho, J.R.S.C.; Aguiar, A.D.C.F. Nutrient use efficiency in alley cropping systems in the Amazonian periphery. *Plant Soil* **2010**, *335*, 363–371. [CrossRef]

38. Taiz, L.; Zeiger, E.; Moller, I.; Murphy, A. *Fisiologia e Desenvolvimento Vegetal*, 6th ed.; Artmed: Porto Alegre, Brazil, 2017; p. 888.

39. Korwar, G.R. Fodder production potential of leucaena hedgerows on an Alfisol and a Vertisol in the semi-arid tropics. In Proceedings of the International Workshop on Nitrogen Fixing Trees for Fodder Production, Morillion, AR, USA, 20–25 March 1995; pp. 146–153.

Article

Management of High-Residue Cover Crops in a Conservation Tillage Organic Vegetable On-Farm Setting

Ted S. Kornecki and Andrew J. Price *

National Soil Dynamics Laboratory, Agricultural Research Service, United States Department of Agriculture, Auburn, AL 36830, USA; ted.kornecki@usda.gov
* Correspondence: price@ars.usda.gov; Tel.: +1-334-887-8596

Received: 16 September 2019; Accepted: 10 October 2019; Published: 15 October 2019

Abstract: A three year on-farm conservation-tillage experiment was initiated in fall of 2008 at Randle Farm LLC, located in Auburn, AL. Our objective was to evaluate and demonstrate implementation of tenable conservation vegetable production practices using high amounts of cover crop residues that reduce soil erosion, improve soil productivity and quality, reduce energy costs, and promote farm profitability. Cereal rye, crimson clover, and a rye and crimson clover mixture were evaluated as cover crops; these were terminated using either a prototype two-stage roller/crimper alone or followed by an application of 2.5 L a.i. ha^{-1} 45% cinnamon (*Cinnamomum verum* L.) oil (cinnamaldehyde, eugenol, eugenol acetate,)/45% clove oil (eugenol, acetyl eugenol, caryophyllene) mixture in the spring prior to crop establishment. A winter fallow conventional tillage system was included for comparison. Watermelons, cantaloupes, and okra then were transplanted into each cover crop and termination treatment combination in mid-May, utilizing a modified transplanter equipped with a custom fitted subsoiling shank and row cleaners to alleviate soil compaction and facilitate transplanting. In all years, all cover crop treatments exceeded 4000 kg ha^{-1} and in 2009 and 2011, exceeded 6000 kg ha^{-1}. At 21 days after termination in 2010 when the slowest termination occurred, higher termination rates were obtained for cereal rye (95% to 96%) followed by lower termination rates for the clover/rye mixture (83% to 85%); the lowest termination rates were obtained for crimson clover (66% to 68%). Commercially available cinnamon/clove oil solution provided little cover crop termination above that provided by a roller crimper alone. Volumetric soil moisture content for rolled/crimped cover crops was consistently higher compared to the conventional system, indicating that flattened and desiccated cover crop residue provided water conservation. In 2010 and 2011, yields for cantaloupe, okra, and watermelons were consistently higher for the conventional system compared with no-till system with cover crops likely due to weed cultivation limitations and insect pressure. Future studies need to focus on weed control and integrated pest management.

Keywords: cinnamon oil; clove oil; cover crop termination; organic agriculture; organic herbicides; roller/crimper

---

1. Introduction

Cover crops are an integral component in conservation agriculture because they provide important benefits that enhance soil quality and plant growth [1]. Recent incentives from United States (U.S) Department of Agriculture (USDA) Natural Resources Conservation Service (NRCS) to utilize mono- and poly-culture cover crops in no-till systems has increased use of these crops in the U.S. To maximize cover crop benefits, they must produce optimum biomass [2]. Commonly used cover crops in the southern United States are cereal rye (*Secale cereale* L.) and crimson clover (*Trifolium incarnatum* L.). Based on historic data, the average cereal rye biomass production level in Alabama is 6000 kg ha^{-1} [3], although cereal rye biomass production can attain 11,100 kg ha^{-1} when growing conditions are optimal [4]. Similar biomass production by crimson clover of 6000 kg ha^{-1} was reported in central Alabama [5], but higher biomass production (7000 kg ha^{-1}) has been also reported [4]. In addition to biomass, crimson clover as a legume can fix nitrogen which is an important alternative to fertilizers as a nitrogen source in an organic production system [6,7]. Major cover crop benefits consist of soil protection from impact of rainfall energy, reduced runoff, decreased soil compaction and increased infiltration [8–12]. Cover crops also provide a physical barrier on the soil surface which inhibits weed germination, emergence and growth [13–16]. In addition to providing a physical barrier, cereal rye possesses allelopathic properties that provide control similar to applying a pre-emergence herbicide [17,18]. Improved soil physical/chemical properties from increasing soil organic carbon, are conditions for better crop growth and sustainable agriculture [19].

Rolling/crimping practices have been used in conservation systems to manage cover crops by flattening and crimping cereal or legumes cover crops [20–22]. Crimping cover crop tissue causes plant injury and accelerates its termination rate [21]. In the southern U.S., terminating cover crops should typically be carried out three weeks prior to planting the cash crop which is similar to normal herbicide burndown recommendations [22,23]. Typically, three weeks after rolling, the termination rate for rye is above 90% when rolling is performed at an optimal growth stage [20–24]. Most agricultural extension services recommend terminating the cover crop at least two weeks prior to planting the cash crop to prevent the cover crop from depleting soil moisture. Hargrove and Frye [6] reported a minimum 14 days from cover crop termination before cash crop planting enable soil water recharge. When winter and early spring months are unusually cold and wet, or too dry, producers must wait longer for rye to reach the optimum growth stage and biomass, thus causing later cash crop planting dates that likely decrease yield potential. A reduction in time between cover crop termination and cash crop planting might also create residue management problems for planting equipment. This is especially critical in vegetable production when delays in transplanting negatively affect growth of plants and yield. On the other hand, warm weather and plentiful rainfall in spring can increase weed pressure and insect populations. Timing is a very important aspect of using cover crops in vegetable production systems to effectively manage nutrient competition, moisture retention, and transplanting success.

Previous research has shown if there is insufficient time between cover crop termination and planting of a cash crop, the cover crop might not completely loose its elasticity, strength, and moisture, making planting difficult due to the possibility of frequent wrapping and accumulation of cover crop residue on planting units, as well as increasing the possibility of hair-pinning [25]. One effective way to reduce the time between terminating cover crops and planting the cash crop is to apply herbicide with rolling operation [22]. In no-till organic production, synthetic herbicides are prohibited, therefore organic herbicides effectiveness must be evaluated at realistic farming conditions.

The objective of this experiment was to evaluate and demonstrate implementation of tenable on-farm conservation vegetable production systems using high amounts of cover crop residues. The experiment evaluated management of three different cover crops in three vegetable crop systems. An organic herbicide was also evaluated for cover crop termination.

2. Materials and Methods

2.1. Experimental Design and Cover Crop Management

A three year on-farm organic no-till experiment was initiated in fall 2008 at Randle Farms LLC, Auburn, AL in the same field. A sandy soil [Cation Exchange Capacity (CEC) = 4.6–9.0 cmolc·kg^{-1}], pH = 5–7) area (0.41 ha) which had not been under crop production for decades, was selected by the producer to conduct a replicated experiment with cover crops. The experiment was a split–split block design with factorial treatment arrangement, with each treatment replicated three times, and the experiment repeated over three years. The main block was cash crop, consisting either of watermelon, cantaloupe, or okra (Figure 1). Within each main block there were four subplot cover crop treatments:

(1) crimson clover, (2) cereal rye (Wrens Abruzzi), (3) a crimson clover + cereal rye mixture, and (4) conventional tilled soil using multiple passes with a disk and field cultivator to prepare the seedbed, with no cover crop. In addition within each cover crop treatment, two cover crop termination treatments were applied to each sub-plot consisting of: (1) rolling/crimping alone, or (2) rolling/crimping followed by an application of 2.5 L a.i. ha^{-1} 45% cinnamon (*Cinnamomum verum* L.) oil (cinnamaldehyde, eugenol, eugenol acetate,)/45% clove oil (eugenol, acetyl eugenol, caryophyllene) mixture (Weed Zap®, JH Biotech, Inc., Ventura, CA, USA) (Figure 1). Plots were 24.4 m long and 1.8 m wide with a 1.8 m alley between each plot.

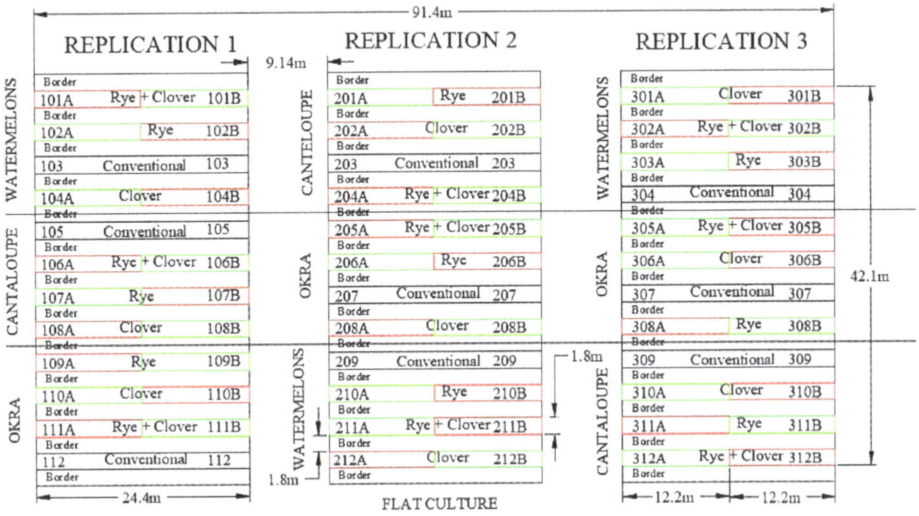

← 91.4m →

REPLICATION 1	REPLICATION 2 (9.14m)	REPLICATION 3
Border	Border	Border
101A Rye + Clover 101B	201A Rye 201B	301A Clover 301B
Border	Border	Border
102A Rye 102B	202A Clover 202B	302A Rye + Clover 302B
Border	Border	Border
103 Conventional 103	203 Conventional 203	303A Rye 303B
Border	Border	Border
104A Clover 104B	204A Rye + Clover 204B	304 Conventional 304
Border	Border	Border
105 Conventional 105	205A Rye + Clover 205B	305A Rye + Clover 305B
Border	Border	Border
106A Rye + Clover 106B	206A Rye 206B	306A Clover 306B
Border	Border	Border
107A Rye 107B	207 Conventional 207	307 Conventional 307
Border	Border	Border
108A Clover 108B	208A Clover 208B	308A Rye 308B
Border	Border	Border
109A Rye 109B	209 Conventional 209	309 Conventional 309
Border	Border	Border
110A Clover 110B	210A Rye 210B	310A Clover 310B
Border	Border	Border
111A Rye + Clover 111B	211A Rye + Clover 211B	311A Rye 311B
Border	Border	Border
112 Conventional 112	212A Clover 212B	312A Rye + Clover 312B
Border	Border	Border

Replication 1 left labels: WATERMELONS, CANTALOUPE, OKRA. Replication 2 labels: CANTELOUPE, OKRA, WATERMELONS. Replication 3 labels: WATERMELONS, OKRA, CANTALOUPE.

Dimensions: 24.4m (Replication 1 width); 1.8m; FLAT CULTURE; 12.2m — 12.2m (Replication 3); 42.1m; 1.8m.

Figure 1. Experiment layout for the on-farm organic no-till experiment.

Cereal rye (var. Wrens Abruzzi, 100 kg ha^{-1}), crimson clover (var. Dixie, 24.5 kg ha^{-1}), and a mixture of rye and crimson clover cover crops were seeded in the fall of 2008, 2009, and 2010 using a no-till Great Plains 606 NT drill with 19 cm row spacing. Cover crop height was assessed, and biomass samples were collected the day before cover crop termination in each spring, dried, and weighed. Cover crops were rolled/crimped mid-April of 2009, 2010, and 2011 using an experimental two-stage 2.4 m wide roller/crimper (Figures 2 and 3) [26] when cereal rye reached the early milk growth stage (Zadoks #77,) [27], which is a desirable termination stage for this species that typically produces the highest biomass [20]. Termination rates were determined using a portable, handheld, active light sensor, (Greenseeker RT100 data collection and mapping unit, NTech Industries, Ukiah, CA, USA). Data were collected continuously by walking with the unit through the length of each plot. With this approach, we hypothesized that dead plants have 0% greenness [28]. In addition, volumetric soil moisture content was measured using time domain reflectometry (TDR) by a FieldScount® TDR 300 (Spectrum Technologies®, Aurora, IL, USA).

Figure 2. Cereal rye crimson clover mixture being rolled/crimped by an experimental patented 1.8 m wide 2-stage roller/crimper.

Figure 3. Side view of the experimental patented two-stage roller/crimper with two adjustable compression springs to set crimping force of the secondary crimping drum.

2.2. Cash Crop Transplanting

Cash crop treatments were established after cover crop termination in mid-May utilizing a modified single row transplanter (RJ Equipment®, Blenheim, ON, Canada N0P 1A0), equipped with a custom-made frame (modifications were made at the National Soil Dynamics Laboratory, Auburn, AL) to which a custom designed slim shank (16 mm thickness) was attached, to alleviate a soil compaction layer (Figure 4). In addition, a spring-loaded fluted coulter and row cleaners (Yetter™, PO BOX 358 109 S. McDonough, Colchester, IL, USA) were used to manage cover crop residue on the soil surface. All plants received irrigation at planting. Drip irrigation was then immediately installed on all plots to provide plants with irrigation, applied four times a day for fifteen minutes until harvest. Hydrolyzed Fish Fertilizer 2-4-1 (Neptune's Harvest, Gloucester, MA, USA) was applied throughout the season through the drip irrigation applied as needed, to meet requirements recommended by the Alabama Cooperative Extension System [29].

Figure 4. (**A**) RJ Equipment® no-till transplanter, with the additional custom designed sub-frame to accommodate a custom subsoiler shank (subsoiler used is pointed by arrow) for a single combined subsoiling and transplanting operation. (**B**) Row cleaners were mounted for cover crop residue management.

Crimson Sweet watermelon (var. *lanatus*), Athena cantaloupe (var. *cantalupensis*), and Clemson Spineless okra (var. *esculentus*) transplants were grown from seed in the greenhouse. Vegetables were hand harvested from all plots and weighed for yield assessment. Field operation timing are presented in Table 1.

Table 1. Field operations at each growing season.

Field Activities	Growing Season		
	2009 (mm/dd/yy)	2010 (mm/dd/yy)	2011 (mm/dd/yy)
Planting cover crops	11/04/08	10/06/09	10/01/10
Terminating cover crops (rolling, ZAP)	04/28/09	04/21/10	04/13/11
Planting watermelons and cantaloupe	06/07/09 (watermelons only)	05/11/10	05/04/11
Planting okra	07/16/09	5/17/10	05/04/11
Harvesting watermelons	08/14/09 to 09/15/09	07/15/10 to 8/16/10	07/13/11 to 08/15/11
Harvesting cantaloupe	X	07/15/10 to 08/16/10	07/13/11 to 08/15/11
Harvesting okra	X	06/22/10 to 08/13/10	06/17/11 to 08/15/11

2.3. Data Analysis

Cover crop termination rates (%) and volumetric soil moisture content (%) were transformed using an arcsine square-root transformation method [30], but this transformation did not result in a change in analysis of variance (ANOVA), thus non-transformed means are presented. Non-transformed cover crop height, cover crop biomass, and cash crop yield data were used in analysis of variance (ANOVA). Cover crop production system, termination system, and their interactions were considered fixed effects. Treatment means were separated using the ANOVA General Linear Model (GLM) procedure; the Fisher's protected least significant differences (LSD) test at the 10% probability level was used [31]. Fixed effect *p* values are reported in Table 2.

Table 2. Fixed effects of (*p* values) for dependent variables utilized in this study.

Source	DF	Cover Crop Height Pr > F	DF	Biomass Pr > F	DF	Cash Crop Yield Cantaloupe Pr > F	Okra Pr > F	Watermelon Pr > F
Year	2	<0.0001	1	<0.0001	1	0.7065	0.2362	0.1760
Cover	3 *	<0.0001	2	<0.0001	3 **	0.2589	0.5492	<0.0001
Rolling Treatment	1	0.1104	1	0.9727	1	0.4634	0.2325	0.8195
Block	2	0.0873	2	0.0949	2	0.3085	0.5114	0.0006
Block * Cover	6	0.4092	6	0.0737	6	0.6538	0.4347	0.0035
Block * Rolling Treatment	2	0.5695	2	0.1829	2	0.6477	0.3740	0.7902

* Degrees of freedom for height include three cover crops (rye, clover, and clover/rye mixture where height of rye and clover were measured separately). Degrees of freedom for biomass indicate biomass for rye, clover, and combined biomass from clover/rye mixture. ** Degrees of freedom for cash crop yield include three cover crops (rye, clover, clover/rye mixture for no-till system, and no-cover crop in conventional tillage system).

3. Results and Discussion

3.1. Cover Crop Height and Biomass Production

Cover crop height was significant for year and species (ANOVA revealed *p* value <0.0001 for both variables). Cereal rye height as monoculture was 161, 138, and 164 cm in 2009, 2010, and 2011 respectively, whereas rye height was decreased in 2009 when mixed with crimson clover (Table 3). The height for crimson clover was 68, 34, and 57 cm alone in 2009, 2010, and 2011 respectively, and did not decrease when mixed with rye. Across three growing seasons, significant differences were detected among cover crops biomass (*p* value for variable YEAR was <0.0001). Similarly, the *p* value for variable COVER was <0.0001. Biomass production averaged over cover crops was highest in 2009 (7753 kg ha^{-1}) followed by lower biomass in 2011 (7130 kg ha^{-1}), and the lowest biomass was recorded in 2010 (4144 kg ha^{-1}). The lowest biomass generated in 2010 was associated with drier field conditions due to a lack of rainfall during cover crop growth period.

Table 3. Height (cm) and cover crop biomass (kg ha^{-1}) on dry basis during three growing seasons.

Cover Crop		Growing Season 2009 Height	Biomass	2010 Height	Biomass	2011 Height	Biomass
Crimson Clover		68.4 c *	7198 *	34.0 b	2956 b	56.5 b	6274 b
Cereal Rye		160.5 a	7940	138 a	4763 a	164.4 a	7052 b
Clover	Clover and Rye Mix	64.8 c	8120	37.9 b	4712 a	60.5 b	8064 a
Rye		153.0 b		132.8 a		162.5 a	
p Value		<0.0001	0.2382	<0.0001	0.0065	<0.0001	0.0151
LSD		5.7	N/S	8.1	1030	4.8	994

* Same lower-case letters within each column represent no difference in height and biomass production.

In 2009, there were no significant difference between biomass production for rye, clover alone, and mixture between these cover crops (*p* value = 0.2382). In 2010, the lowest cover crop biomass was generated by crimson clover alone (2956) compared with significantly higher biomass for cereal rye and clover/rye mixture (4763 and 4712 kg ha^{-1}, respectively). In 2011, significantly higher biomass was generated by clover/rye mix (8064 kg ha^{-1}) compared with lower biomass for clover (6274 kg ha^{-1}) and cereal rye 7052 (kg ha^{-1}) alone.

3.2. Cover Crop Termination

Cereal rye had significantly higher termination rates compared to clover/rye mix and clover alone across three growing seasons where the termination rates were averaged over one, two, and three weeks after treatment application. Overall, the lowest termination rates were observed for crimson clover (Table 4). The results show that applying organic herbicide ZAP in addition to mechanical termination by rolling/crimping did not improve termination rates for single species cover crops (rye and clover) and clover/rye mixture for all three growing seasons.

Table 4. Three-year average termination rates comparison between rolled only treatment and roller with ZAP application treatment.

Cover Crop	Treatment	Termination Rate (%)
Cereal Rye	Rolled/crimped only	75.0 a
	Rolled/crimped with ZAP	75.5 a
Crimson Clover	Rolled/crimped only	64.1 c
	Rolled/crimped with ZAP	65.1 c
Cereal Rye and Crimson Clover Mixture	Rolled/crimped only	71.6 b
	Rolled/crimped with ZAP	71.1 b
GLM Procedure Results	*p* value for rolling treatment	<0.0001
	LSD	3.0

* Same lower-case letters within last column represent no difference in cover crop termination rates.

Overall rye and mixture were easier to terminate (above 90% three weeks after rolling) (Table 5). Clover alone, because of its very low height in 2010, was unable to engage with the roller's crimping bars and resulted in lower termination three weeks after rolling (68%).

Table 5. Cover crop termination rates during 2009, 2010, and 2011 growing seasons.

Cover Crop	TRT	Growing Season								
		2009			2010			2011		
		Days after Treatment Application								
		7	14	21	7	14	21	7	14	21
Cereal Rye	Rolling	56	84 a *	100 a	59 a	63 b	95 a	46 b	74	96
	Rolling + ZAP	51	83 a	100 a	57 a	71 a	96 a	55 a	71	96
Crimson Clover	Rolling	50	66 c	91 b	65 a	52 c	68 c	29 d	64	92
	Rolling + ZAP	54	72 bc	93 b	62 a	52 c	66 c	32 cd	66	91
Clover/Rye Mixture	Rolling	55	74 b	100 a	63 a	64 b	83 b	41 b	69	96
	Rolling + ZAP	59	77 ab	100 a	60 a	60 b	85 b	38 bc	67	93
p Value		0.37	0.0004	<0.0001	0.0208	<0.0001	<0.0001	<0.0001	0.37	0.1042
LSD		N/S	6.8	2.6	5.4	6.8	6.3	8.6	N/S	N/S

* Same lower-case letters within each column represent no difference in cover crop termination rates.

At 14 days after rolling, significantly higher termination rates of 83% and 84% were associated with rolling rye and rolling rye with a supplemental ZAP application, respectively. However, these termination rates were not different than for rolling clover with ZAP (77%). Termination rates for crimson clover were significantly lower (66%) for rolling only treatment and slightly higher (72%) for rolling with ZAP, but there were no differences between no-ZAP and ZAP treatments for clover. Data clearly suggest that adding ZAP to rolling did not increase termination for both mono and mix cover crops. At 21 days after rolling and chemical treatment of ZAP, higher termination rates were generated for rye (100%) and clover/rye mix (100%) with or without ZAP treatment, compared with lower termination rates for clover cover crop (91% for rolling only and 93% for rolling with ZAP).

At 14 days after rolling treatment, higher termination rate for cereal rye was associated with rolling with ZAP treatment compared to roller alone (71% vs. 63%). For crimson clover, there was no difference in termination rates between roller only and roller plus ZAP treatments. However, no difference in termination rates for rye and clover mixture were found between rolling only and rolling with ZAP treatment. Overall, termination rates for crimson clover (52%) were lower for both rolling treatments compared with cereal rye and clover/rye mixture. At 21 days there was no difference between rolling only and rolling with ZAP for each cover crop. Higher termination rates were obtained for cereal rye (95% to 96%) followed by lower termination rates for clover/rye mixture (83% to 85%) and the lowest termination rates were obtained for crimson clover (66% to 68%).

In 2011 at seven days after rolling treatment application, higher termination rate of 55% was obtained for rolled rye and ZAP followed by lower termination for rolled rye (46%), rolled clover/rye mix (41%), and rolling with ZAP (38%). As in previous growing seasons, the lowest termination rates were observed for crimson clover having 29% for rolled only treatment and 32% for rolling with ZAP treatment, without statistical differences between these treatments. At 14 days after rolling, there were no differences in termination among cover crops and rolling treatments ranging from 64% (crimson clover with roller only) to 74% (for cereal rye roller only treatment). Similarly, at 21 days after rolling, no differences in termination rates among cover crops and rolling treatments were found having termination rates ranging from 91% for crimson clover (rolling with ZAP) to 96% for both cereal rye (rolling with and without ZAP) and clover/rye mixture (for rolling only treatment).

Termination data from three growing seasons clearly indicates that adding organic ZAP herbicide with rolling did not increase termination rates for all cover crops. Therefore, based on this study, spending money for Weed ZAP application is not recommended due to expense and ineffectiveness.

3.3. Volumetric Soil Moisture Content

Across three growing seasons, rolling treatments had a significant effect on volumetric soil moisture content (p value < 0.0001) (Table 6). The higher volumetric moisture content (VMC) was observed for rolled rye + ZAP (6.6%), rolled clover/rye mix (6.5%), rolled clover/rye mix + ZAP (6.2%), and rolled rye (6.1%) without significant differences among these treatments. Lower volumetric soil moisture content was observed for rolled crimson clover (5.7%) and rolled crimson clover + ZAP (5.7%) but not significantly different than VMC obtained for rolled clover/rye mix + ZAP and rolled rye alone. The lowest VMC of 4.4% was recorded for conventional culture (rototilled soil without cover crops). These results indicate that rolled residue provided consistently higher soil moisture during three weeks of evaluation in addition to protecting the soil surface from water and wind erosion.

Table 6. Volumetric soil moisture content during 2009, 2010, and 2011 growing seasons.

Cover	TRT	Growing Season								
		2009			2010			2011		
		Days after Treatment Application								
		7	14	21	7	14	21	7	14	21
Cereal Rye	Rolling	9.1 abc	8.3 a *	8.2	4.6 ab	7.1 a	3.0 a	6.7 ab	2.4 ab	6.0
	Rolling + ZAP	9.4 ab	8.4 a	8.5	4.8 ab	7.9 a	3.0 a	6.8 a	3.0 a	7.3
Crimson Clover	Rolling	9.2 abc	7.2 b	7.5	4.6 ab	6.8 a	1.7 c	6.5 ab	1.6 c	6.7
	Rolling + ZAP	8.9 bc	7.2 b	8.1	4.3 b	7.1 a	1.8 bc	6.8 a	1.5 c	5.6
Clover/Rye Mixture	Rolling	10.2 a	9.0 a	8.4	5.2 a	8.0 a	3.2 a	6.6 ab	1.7 bc	6.3
	Rolling + ZAP	9.7 ab	8.8 a	8.2	4.9 ab	7.5 a	2.7 ab	5.9 b	1.8 bc	6.3
No Cover, No Rolling		7.8 d	6.2 c	7.0	2.2 c	5.5 b	0.8 d	3.1 c	1.2 c	6.0
p Value		0.0093	<0.0001	0.1376	<0.0001	0.0063	<0.0001	<0.0001	0.0083	0.3682
LSD		1.16	1.02	N/S	0.71	1.31	0.89	0.83	0.86	N/S

* Same lower-case letters within each column represent no difference in volumetric soil moisture content.

In 2009, at seven days after rolling and ZAP application, VMC for rolled cover crops with or without ZAP treatment ranged from 8.9% (rolled crimson clover + ZAP) to 10.2% (rolled clover/rye mix) and was significantly higher than for conventional operation (7.8%). Except for crimson clover (rolling + ZAP treatment) and clover/rye mix, no significant differences in VMC were detected among other rolled residue treatments indicating that rolling effectively preserved soil water. At 14 days after rolling, significantly higher VMC (8.3% to 9%) was observed for cereal rye and clover/rye mix for rolled and rolled plus ZAP applications compared to crimson clover (rolled and rolled–ZAP with 7.2% VMC) and the lowest VMC of 6.2% for conventional tillage without cover crops most likely due to higher water evaporation from the bare soil. Lower VMC for crimson clover might be associated with lesser soil coverage due to numerically lower biomass production for clover compared to rye and clover/rye mix. At 21 days after rolling, there were no significant differences in VMC among all treatments with 7% VMC for conventional plots (numerically lowest), and 7.5% to 8.5% for rolled treatments.

In 2010, compared to the previous growing season, the volumetric soil water in 2010 during the three weeks of the evaluation was significantly lower due to lack of rainfall. At seven days the VMC was 4.3% for rolled clover plus ZAP which was significantly lower than rolled clover/rye mix with VMC of 5.2%. All other rolling treatments for cereal rye (rolled only and rolled with ZAP), rolled clover and mixture (rolled only and rolled with ZAP) had VMC from 4.6% to 4.9% without significant difference among these treatments. In contrast conventional plots had significantly lower VMC of 2.2% indicating higher water evaporation compared with the covered soil surface by cover crop residue. At 14 days after rolling, there were no significant differences in VMC among all rolled and ZAP treatments for all covers ranging from 6.8% to 8.0%. The conventional plots had significantly lower VMC of 5.5% compared to cover crops treatments. At 21 days after rolling, because of a lack of rainfall, the VMC for all treatments was significantly lower compared with seven-day and 14-day evaluations. The VMC readings of 3.0% for cereal rye (rolled and rolled with ZAP) and 2.7% to 3.2% for clover/rye mix (rolled and rolled with ZAP) was higher compared to lower VMC for rolled clover (1.7%) and rolled with ZAP (1.8%). These lower VMC levels were most likely associated with unusually low production of clover biomass, approximately 1.8 metric tons less than for cereal rye and clover/rye mix. The conventional plots without covers had very low VMC of 0.8% indicating severe drought conditions.

In 2011, at seven days after rolling and ZAP application, no significant differences in VMC were detected among rolled clover and cereal rye (with and without ZAP) and for rolled only clover/rye mix. Lower VMC of 5.9% was obtained for rolled clover/rye mix plus ZAP compared with higher VMC for rolled clover and rye plus ZAP (6.8%). The conventional system without cover crop had significantly lower VMC of 3.1% compared to two times higher VMC for rolled cover crop treatments. At 14 days after rolling, VMC for all cover crops was unusually low due to a drought and higher temperatures. The lowest VMC was obtained for conventional system (1.2%), followed by crimson clover (1.6% for rolling and 1.5% for rolling with ZAP) and by clover/rye mixture (1.7% for rolling only and 1.8% for rolling with ZAP). VMC for cereal rye was higher (3.0% for rolling with ZAP). VMC for rolling only was 2.4% but not statistically different than VMC for clover/rye mixture. At 21 days after rolling, compared with 14 days, VMC was higher for all cover crops and rolling treatments due to a rainfall occurring between 14 and 21 days after rolling. This rainfall event elevated VMC ranging from 6.0% to 7.3% without statistical difference in VMC among cover crops and rolling treatments. Generally, across all growing seasons, rolled cover crops had higher VMC compared to conventional system indicating water conservation as one of the major benefits that cover crops provide.

3.4. Cash Crop Yield

In 2009, there were many problems associated with no-till systems when commercial pesticides were not used. Large insect populations (grasshoppers and squash bugs) severely hindered crop establishment and high weed pressure reduced yields. The 2009 growing season served as a learning process to resolve pest problems by using organic compounds and mechanical methods. Organic insecticides were used in late 2009 including Hot Pepper Wax and Pyganic E.C. 5.0. for insect control.

1. Cantaloupe. There were no significant differences in cantaloupe yield between two growing seasons (p value = 0.6995). In 2010 the yield had higher numerical value of 3037 kg ha^{-1} compared to 2710 kg ha^{-1} in 2011 (Table 7). Across two growing seasons, there was no significant difference in cantaloupe yield among three blocks (p value = 0.2906). In addition, no significant differences in the yield were present between rolling alone and rolling + ZAP, although the rolling + ZAP treatment had a higher numerical yield of 3193 kg ha^{-1} compared to lower yield of 2554 kg ha^{-1} for rolling only treatments. In 2010 cover crops did not influence cantaloupe yield (Table 7) ranging from 2382 kg ha^{-1} for cereal rye to 4126 kg ha^{-1} for conventional plots. In contrast, in 2011 growing season cover crop did have an effect on the yield with a lower 606 kg ha^{-1} for cereal rye and similar yields of 3823 kg ha^{-1} for conventional system, 2746 kg ha^{-1} for clover and 3665 kg ha^{-1} for clover/rye mixture. However, no differences found between rolling only and rolling with ZAP treatments (p value = 0.7064). Likewise, rolling treatment across two growing seasons, did not have an influence on yield (p value = 0.4521) generating cantaloupe yield of 3194 kg ha^{-1} for rolled + ZAP compared to 2554 kg ha^{-1} for rolled only treatment.

2. Okra. There was a significant difference in okra yield between two growing seasons (p value = 0.0051); the yield in 2010 was 1207 kg ha^{-1} compared to 1603 kg ha^{-1} in 2011 (Table 8). Moreover, across two growing seasons, significant difference in okra yield existed among the three blocks (p value = 0.0001) and among cover crops (p value < 0.0001). In 2010 growing season, cover crops did have an effect on the okra yield, with lowest yield of 525 kg ha^{-1} for crimson clover/cereal rye mix, 639 kg ha^{-1} for crimson clover without significant difference between these cover crops, and higher yield of 839 kg ha^{-1} was found for cereal rye, and the significantly higher yield of 2823 kg ha^{-1} was obtained for conventional system compared to all cover crops. Similarly, in 2011 growing season, cover crops did influence okra yield (Table 8) ranging from the lowest yield of 935 kg ha^{-1} for cereal rye, followed by 1146 kg ha^{-1} for crimson clover (without significant difference between these cover crops) and the higher okra yield of 1472 kg ha^{-1} for the clover/rye mix. Compared to the cover crops used, conventional plots generated significantly higher okra yield of 2860 kg ha^{-1}. In addition, across two growing seasons, rolling treatment did not have an influence on yield (p value = 0.9551).

3. Watermelon. No significant difference existed in watermelon yield between two growing seasons (p value = 0.1760), but in 2010 the watermelon yield had higher numerical value of 6257 kg ha^{-1} compared to 4827 kg ha^{-1} in 2011 (Table 9). Across two growing seasons cover crops influenced watermelon yield (p value < 0.0001). Significant difference in watermelon yield was observed among blocks (p value = 0.0006) indicating high variability in the yield among three blocks. In addition, no significant differences in watermelon yield were found between rolling alone (5661 kg ha^{-1}) and rolling + ZAP (5424 kg ha^{-1}). In 2010 cover crops did influence watermelon yield (Table 9) ranging from 2382 kg ha^{-1} for cereal rye to 4126 kg ha^{-1} for conventional plots. Similarly, in 2011 growing season, cover crop did have an effect on the yield from lowest of 2258 kg ha^{-1} for cereal rye to the highest 8315 kg ha^{-1} for conventional system. There was no difference in watermelon yield in 2011 between crimson clover/rye mix (5211 kg ha^{-1}) and crimson clover (3525 kg ha^{-1}), but the watermelon yield for crimson clover was not statistically different than for cereal rye. The 2011 growing season showed, similar to 2010, that rolling treatment did not have an influence on yield (p value = 0.5017) generating watermelon yield of 5147 kg ha^{-1} for rolled compared to numerically lower yield of 4508 kg ha^{-1} for rolled + ZAP treatment with the conventional system having a significantly higher yield of 8315 kg ha^{-1}.

Table 7. Cantaloupe yield (kg ha^{-1}) with respect to cover crop treatments in 2010 and 2011 growing seasons.

Cover Crop	Growing Season		Average Yield over Two Growing Seasons
	2010	2011	
Cereal Rye	2382	606 b	1495
Crimson Clover	3461	2746 a	3103
Clover/Rye mixture	2179	3665 a	2922
Conventional	4126	3823 a	3974
p Value	0.7285	0.0117	0.2380
LSD at α = 0.10	N/S	1454 kg ha^{-1}	N/S
Yield Averaged over Cover Crops	3037	2710	p value = 0.6995
Rolling Treatment: Rolling Only	2505	2602	2554
Rolling Treatment: Rolling with ZAP	3569	2818	3194
p Value	0.4627	0.7064	0.4521

* Same lower-case letters in each column represent no significant difference in cantaloupe yield among cover crops.

Table 8. Okra yield (kg ha^{-1}) with respect to cover crop treatments in 2010 and 2011 growing seasons.

Cover Crop	Growing Season		Average Yield over Two Growing Seasons for Each Cover Crop
	2010	2011	
Cereal Rye	839 b	935 c	887 b
Crimson Clover	639 c	1146 bc	893 b
Clover/Rye Mixture	525 c	1472 b	999 b
Conventional	2823 a	2860 a	2842 a
p value	<0.0001	<0.0001	<0.0001
LSD at α = 0.10	155	442	311
Yield Averaged over Cover Crops	1207 B **	1603 A	p value = 0.0051
Rolling Treatment: Rolling Only	1196	1621	1408
Rolling Treatment: Rolling with ZAP	1217	1585	1401
p Value	0.7252	0.8359	0.9551

* Same lower-case letters in each column represent no significant difference in okra yield among cover crops. ** Same upper-case letters in the row indicate no significant difference in okra yield (growing seasons).

Table 9. Watermelon yield (kg ha^{-1}) with respect to cover crop treatments in 2010 and 2011 growing seasons.

Cover Crop	Growing Season		Average Yield over Two Growing Seasons
	2010	2011	
Cereal Rye	3493 b *	2258 c *	2876 c
Crimson Clover	5414 b	3525 bc	4470 b
Clover/Rye Mixture	2877 b	5211 b	4044 b
Conventional	13,244 a	8315 a	10780 a
p Value	0.0004	0.0060	<0.0001
LSD at $\alpha = 0.10$	2979	2365	2481
Yield Averaged over Cover Crops	6257	4827	*p* value = 0.1760
Rolling Treatment: Rolling Only	6174	5147	5661
Rolling Treatment: Rolling with ZAP	6340	4508	5424
p Value	0.8888	0.5017	0.8195

* Same lower-case letters in each column represent no significant difference in watermelon yield among cover crops.

Results from this experiment and previous research indicates that yields in no-till systems using cover crops tend to be lower in the first growing season compared to conventional systems [32].

4. Conclusions

This work presents findings for a vegetable no-till system using cover crops under weather and soil conditions of Alabama and provides guidance for adoption of similar conservation systems in the Southeast. Across three growing seasons, cover crop termination rates by roller/crimper alone were consistently as good as rolling/crimping with supplemental application of organic herbicide Weed ZAP. Termination results indicate that Weed ZAP was ineffective in speeding up the termination process, therefore extra costs incurred for purchasing this organic product and cost of application was added to overall cost of termination without tangible results, therefore is not recommended for cover crop termination. Volumetric soil moisture content for rolled/crimped cover crops was consistently higher compared to the conventional system, indicating that flattened and desiccated cover crop residue provided water conservation. In contrast, in the conventional system, soil was exposed without coverage resulting in lower volumetric soil moisture content. Since this field experiment was focused on a no-till organic system using cover crops, non-tillage operation for plots with planted cover crops, there were many problems encountered in the first year of experiment (2009 growing season). In 2010 and 2011, yields for cantaloupe, okra, and watermelons were consistently higher for the conventional system compared with the no-till system with cover crops. Previous research indicates that yields in no-till systems using cover crops tend to be lower in the first growing season compared to conventional systems [4]. One reason for reduced cash crop yields may be related to lower N availability to cash crops due to N demands by microbes during decomposition of cover crop residue. Also, increased weed population (due to cultivation limitations) and increased insect populations (due to residual habitat) may have hindered yield potential in the no-till system. This study was conducted on-farm under realistic farm conditions and the results indicate that major fluctuations in yield of cash crops were the result of inadequate weed and insect control from use of less effective organic products than synthetic pesticides. Future studies need to be focused on weed control by generating optimum cover crop residues covering soil surface providing an effective mulch barrier to prevent weed emergence as well as an integrated pest management plan to decrease yield impact of pest insects in these systems.

Author Contributions: T.S.K. and A.J.P. collaborated on experimental conceptualization, experimental investigation, contributing to resources, statistical analysis, and writing—original draft preparation.

Funding: This Conservation Innovation Grant (CIG) on-farm project was conducted with a financial support from the USDA-NRCS, Auburn University and USDA-Agricultural Research Service (ARS), National Soil Dynamics Lab.

Acknowledgments: The authors wish to acknowledge Corey Kichler, agricultural engineer, Trent Morton, biological technician and Kirk Iversen, project administrator for their personal involvement in cover crop termination, treatments application, planting, harvesting of cash crops, and for creative ideas such as insect control and maintaining constant communication with the producer Frank Randle.

Conflicts of Interest: The authors declare no conflict of interest.

References

1. Crossland, M.; Fradgley, N.; Henry Creissen, H.; Sally Howlett, S.; Baresel, P.; Finckh, M.; Robbie Girling, R. An online toolbox for cover crops and living mulches. *Asp. Appl. Biol.* **2015**, *129*, 1–6, Getting the Most out of Cover Crops. Available online: orgprints.org/29405/1/1%20FRADGLEY.PDF (accessed on 15 October 2019).

2. Brady, N.C.; Weil, R.R. *The Nature and Properties of Soils*, 12th ed.; Prentice-Hall, Inc.: Upper Saddle River, NJ, USA, 1999; p. 881.

3. Reiter, M.S.; Reeves, D.W.; Burmester, C.H.; Torbert, H.A. Cotton nitrogen management in a high-residue conservation system: Cover crop fertilization. *Soil Sci. Soc. Am. J.* **2008**, *72*, 1321–1329. [CrossRef]

4. Bowen, G.; Shirley, C.; Cramer, C. *Managing Cover Crops Profitably*, 3rd ed.; Sustainable Agriculture Network Handbook Series, Book 9; National Agricultural Library: Beltsville, MD, USA, 2012; p. 244.

5. Kornecki, T.S.; Price, A.J.; Balkcom, K.S. Cotton population and yield following different cover crops termination practices in an Alabama no-till system. *J. Cotton Sci.* **2015**, *19*, 375–386.

6. Hargrove, W.L.; Frye, W.W. The need for legume cover crops in conservation tillage production. In *The Role of Legumes in Conservation Tillage Systems*; Power, J.F., Ed.; Soil Conservation Society of America: Ankeny, IA, USA, 1987; pp. 1–5.

7. Hubbell, D.H.; Sartain, J.B. *Legumes—A Possible Alternative to Fertilizer Nitrogen*; Florida Cooperative Extension Service Circ. SL-9; University of Florida: Gainesville, FL, USA, 1980.

8. Kern, J.S.; Johnson, M.G. Conservation tillage impacts on national soils and atmospheric carbon levels. *Soil Sci. Soc. Am. J.* **1993**, *57*, 200–210. [CrossRef]

9. McGregor, K.C.; Mutchler, C.K. Soil loss from conservation tillage for sorghum. *Trans. ASAE* **1992**, *35*, 1841–1845. [CrossRef]

10. Reeves, D.W. Cover crops and rotations. In *Advances in Soil Science: Crops Residue Management*; Hatfield, J.L., Stewart, B.A., Eds.; Lewis Publ.: Boca Raton, FL, USA, 1994; pp. 125–172.

11. Raper, R.L.; Reeves, D.W.; Burmester, C.H.; Schwab, E.B. Tillage depth, tillage timing, and cover crop effects on cotton yield, soil strength, and tillage energy requirements. *Appl. Eng. Agric.* **2000**, *16*, 379–385. [CrossRef]

12. Raper, R.L.; Reeves, D.W.; Schwab, E.B.; Burmester, C.H. Reducing soil compaction of Tennessee Valley soils in conservation tillage systems. *J. Cotton Sci.* **2000**, *4*, 84–90.

13. Creamer, N.G.; Bennett, M.A.; Stinner, B.R.; Cardina, J.; Regnier, E.E. Mechanisms of weed suppression in Cover Crop-based production systems. *HortScience* **1996**, *31*, 410–413. [CrossRef]

14. Teasdale, J.R.; Mohler, C.R. The quantitative relationship between weed emergence and the physical properties of mulches. *Weed Sci.* **2000**, *48*, 385–392. [CrossRef]

15. Price, A.J.; Korres, N.; Norsworthy, J.S.; Li, S. Influence of a cereal rye cover crop and conservation tillage on the critical weed free period in cotton. *Weed Technol.* **2018**, *32*, 683–690. [CrossRef]

16. Price, A.J.; Williams, J.; Duzy, L.; McElroy, S.; Guertal, B.; Li, S. Effects of integrated polyethylene and cover crop mulch, conservation tillage, and herbicide application on weed control, yield, and economic returns in watermelon. *Weed Technol.* **2018**, *32*, 623–632. [CrossRef]

17. Barnes, J.P.; Putnam, A.R. Rye residues contribute weed suppression in no-tillage cropping systems. *J. Chem. Ecol.* **1983**, *9*, 1045–1057. [CrossRef]

18. Hoffman, L.M.; Weston, L.A.; Snyder, J.C.; Reginer, E.E. Alleopatic influence of germinating seeds and seedlings of cover crops and weed species. *Weed Sci.* **1996**, *44*, 579–584. [CrossRef]

19. Chan, Y. *Increasing Soil Organic Carbon of Agricultural Land*; PRIMEFACT New South Wales (NSW) Government, Department of Primary Industries, 2008; p. 735. Available online: http://www.dpi.nsw.gov.au/__data/assets/pdf_file/0003/210756/Increasing-soil-organic-carbon.pdf (accessed on 15 October 2019).

20. Ashford, D.L.; Reeves, D.W. Use of a mechanical roller crimper as an alternative kill method for cover crop. *Am. J. Altern. Agric.* **2003**, *18*, 37–45. [CrossRef]

21. Kornecki, T.S.; Price, A.J.; Raper, R.L.; Bergtold, J.S. Effectiveness of different herbicide applicators mounted on a roller/crimper for accelerated rye cover crop termination. *Appl. Eng. Agric.* **2009**, *25*, 819–826. [CrossRef]

22. Price, A.J.; Arriaga, F.J.; Raper, R.L.; Balkcom, K.S.; Kornecki, T.S.; Reeves, D.W. Comparison of mechanical and chemical winter cereal cover crop termination systems and cotton yield in conservation agriculture. *Cotton Sci.* **2009**, *13*, 238–245.

23. Reeves, D.W. A Brazilian model for no-tillage cotton production adapted to the southeastern USA. In Proceedings of the II World Congress on Conservation Agriculture—Producing in Harmony with Nature, Iguassu Falls, Paraná, Brazil, 11–15 August 2003; pp. 372–374.

24. Kornecki, T.S.; Price, A.J.; Raper, R.L. Performance of Different Roller Designs in Terminating Rye Cover Crop and Reducing Vibration. *Appl. Eng. Agric.* **2006**, *22*, 633–641. [CrossRef]

25. Kornecki, T.S.; Arriaga, F.J.; Price, A.J.; Balkcom, K.S. Effects of different residue management methods on cotton establishment and yield in a no-till system. *Appl. Eng. Agric.* **2012**, *28*, 787–794. [CrossRef]

26. Kornecki, T.S. Multistage Crop Roller. U.S. Patent Number 7,987,917B, 2 August 2011.

27. Zadoks, J.C.; Chang, T.T.; Konzak, C.F. A decimal code for the growth stages of cereals. *Weed Res.* **1974**, *14*, 415–421. [CrossRef]

28. Kornecki, T.S.; Arriaga, F.J.; Price, A.J. Evaluation of visual and non-visible light active sensors methods to assess termination rates of cover crops. *TASABE* **2012**, *55*, 733–741. [CrossRef]

29. Anonymous. *Nutrient Recommendation Tables for Alabama Crops*; Agronomy and Soils Departmental Series No. 324; Alabama A&M University: Huntsville, AL, USA, 2012.

30. Steel, R.G.; Torrie, J.H. *Principles and Procedures of Statistics*; McGraw: New York, NY, USA, 1960.

31. SAS. *SAS User's Guide: Statistics. Proprietary Software Release 9*; SAS Institute: Cary, NC, USA, 2013.

32. Johnson, T.J.; Kaspar, T.C.; Kohler, K.A.; Corak, S.J.; Logsdon, S.D. Oat and rye over-seeded into soybean as fall cover crops in the upper Midwest. *J. Soil Water Conserv.* **1998**, *53*, 276–279.

Article

Processing Tomato–Durum Wheat Rotation under Integrated, Organic and Mulch-Based No-Tillage Organic Systems: Yield, N Balance and N Loss

Giacomo Tosti *, Paolo Benincasa, Michela Farneselli, Marcello Guiducci, Andrea Onofri and Francesco Tei

Department of Agricultural, Food and Environmental Sciences (DSA3), University of Perugia, Borgo XX Giugno 74, 06121 Perugia, Italy; paolo.benincasa@unipg.it (P.B.); michela.farneselli@unipg.it (M.F.); marcello.guiducci@unipg.it (M.G.); andrea.onofri@unipg.it (A.O.); francesco.tei@unipg.it (F.T.)
* Correspondence: giacomo.tosti@gmail.com; Tel.: +39-075-585-6333

Received: 9 October 2019; Accepted: 4 November 2019; Published: 6 November 2019

Abstract: In a 4-year study, the biannual crop rotation processing tomato–durum wheat was applied to three cropping systems: (i) an innovative organic coupled with no-tillage (ORG+) where an autumn-sown cover crop was terminated by roller-crimping and then followed by the direct transplantation of processing tomato onto the death-mulch cover; (ii) a traditional organic (ORG) with autumn-sown cover crop that was green manured and followed by processing tomato; and (iii) a conventional integrated low-input (INT) with bare soil during the fall–winter period prior to the processing tomato. N balance, yield and N leaching losses were determined. Innovative cropping techniques such as wheat–faba bean temporary intercropping and the direct transplantation of processing tomato into roll-crimped cover crop biomass were implemented in ORG+; the experiment was aimed at: (i) quantifying the N leaching losses; (ii) assessing the effect of N management on the yield and N utilization; and (iii) comparing the cropping system outputs (yield) in relation to extra-farm N sources (i.e., N coming from organic or synthetic fertilizers acquired from the market) and N losses. The effects of such innovations on important agroecological services such as yield and N recycling were assessed compared to those supplied by the other cropping systems. Independently from the soil management strategy (no till or inversion tillage), cover crops were found to be the key factor for increasing the internal N recycling of the agroecosystems and ORG+ needs a substantial improvement in terms of provisioning services (i.e., yield).

Keywords: cover crops; mulch-based system; N leaching; no-till organic system; intercropping; ecological intensification

---

1. Introduction

Several studies have pointed out the urgent requirement to reduce the impact of the food system on the environment, and such a challenge has to be faced in the framework of climate change [1] and the increasing world population [2]. Complex strategies involving agricultural, social, economic and political components at local, national and global scale are needed [3].

Focusing on the farming system, an increasing importance has been attributed to a number of agroecological services (other than yield) that could be supplied and/or enhanced by implementing appropriate technical choices in cropping system management [4]. Among the most interesting practices, the introduction of cover crops (CCs) in the crop rotation represents a key strategy to ensure several agroecosystem services [5]. CCs are particularly important when dealing with organic farming as they are a crucial element for fertility management and weed control [6,7]. In organic farming systems, CCs are usually terminated before the establishment of the following cash crop by cutting and chopping the plant biomass and incorporating it into the soil (i.e., via green manuring [6]). An emerging alternative to green manuring, that aims at reducing the drawbacks related to the intensive soil tillage (i.e., mainly nonrenewable fuel consumption, soil organic matter and soil biodiversity decrease), is represented by the cover crop mulch-based no-tillage management (MBNT, [8]). In MBNT, the CC is mechanically terminated by one or more roller-crimper passages, thus the devitalized biomass acts as a soil-anchored mulch where the following cash crop is directly sown/transplanted. The adoption of MBNT practice in organic cropping systems has the potential to merge the environmental benefits of no-tillage and organic farming [8].

A small volume of research on this topic is available. Nonetheless, several advantages have emerged: the improvement of soil biodiversity [9], the increase in water and nutrient availability [10], increase in carbon sequestration [11] and the reduction of greenhouse gas emission [12]. In contrast, weed competition (from both volunteer plants and CC regrowth) has emerged to be the most critical challenge to be faced in the adoption of MBNT management in organic systems [13,14]. Beside such challenge, to our present knowledge, information is lacking on N balance and N use efficiency in MBNT organic systems, even if it is well known that such issues are of paramount importance [15]. In particular, the N balance of a given crop rotation is greatly influenced by the N leaching loss which in turn determines the extent of the overall N self-sufficiency [16]. The present study is based on a 4-year experiment, where the same cash crop rotation was applied to three cropping systems, at increasing ecological intensification: a conventional integrated system (INT) with bare soil during the fall–winter period prior to the processing tomato; an organic system with autumn-sown cover crop and traditional inversion tillage (ORG); and an innovative MBNT organic systems (ORG+) where processing tomato was directly transplanted onto the death-mulch cover obtained by roller-crimping the cover crop biomass.

The research was aimed at:

Quantifying the N leaching loss occurring in the three cropping systems in relation to the management strategies.

Assessing the effect of N management (i.e., cover crop termination technique and fertilization strategies) on the yield and N utilization efficiency of the cash crops.

Comparing the cropping system outputs (yield) in relation to extra-farm N sources and N losses.

2. Materials and Methods

2.1. Experimental Site and Management of the Cropping Systems

Field experiments were carried out in four consecutive years (2013/14, 2014/15, 2015/16, and 2016/17) at the experimental station (FieldLab) of the Department of Agricultural, Food and Environmental Sciences of the University of Perugia, Italy. The FieldLab is located in the Tiber river alluvial plain at 42.956°N, 12.376°E, 163 m asl. The soil is a typical Fluventic Haplustept clay-loam (20% sand, 46% silt and 34% clay, 1.4 Mg m^{-3} bulk density), sub-alkaline (pH$_{H2O}$ = 7.8), poor in organic matter (12 g om kg^{-1}, C/N ratio = 11) and in extractable phosphorus (29.9 mg P$_2$O$_5$ kg^{-1}, Olsen method) and rich in exchangeable potassium (258 mg K$_2$O kg^{-1}, int. method).

During the four experimental years, two cycles of the same two-year rotation involving durum wheat (*Triticum durum* Desf., cv Dylan) and processing tomato (*Solanum lycopersicum* L., cv PS1296) were carried out. The same rotation was applied to three cropping systems (treatments) following an increased ecological intensification: conventional integrated (INT), traditional organic (ORG) and innovative organic (ORG+) where a cover crop mulch-based no-tillage system was implemented. Both cash crops were present each year on two adjacent fields (A and B) and they were switched every year from field A to field B. Each field was divided into two blocks where the three treatments were randomly allocated. The dates when all the agronomic operations took place were recorded across the 4-year period (Table S1). The plot size was 540 m^2. The weather data during the whole growing season were obtained from an automatic meteorological station inside the experimental site.

2.1.1. Processing Tomato

In ORG and ORG+, processing tomato was preceded by an autumn-sown cover crop of field pea (*Pisum arvense* L., cv Arkta) and barley (*Hordeum vulgare* L., cv Amyllis) in mixture (barley at 25% + pea at 75% of their ordinary full sowing rates, i.e., 100 and 75 seeds m^{-2}, respectively) while, in INT, the soil was left bare and weed-free (by mechanical control). Cover crop termination was carried out traditionally in ORG: aboveground biomass of the mixture was mowed, finely chopped (0.02–0.1 m) and immediately incorporated into the soil (0.2 m depth) by a rotary cultivator equipped with tines and a back-roller. In ORG+, cover crop biomass was roll-crimped and left on the soil surface as dead mulch. Cover crop termination in ORG+ was generally postponed compared to ORG, because of the slower plant development in the conservative system (Table S1). The processing tomato was transplanted at 3.3 plants m^{-2} into single rows 1 m apart by a standard machinery in the INT and ORG and a no-till direct transplanter in ORG+.

All systems received a fertilization of 150 kg N ha^{-1}, which was distributed by means of fertigation (details on rate, scheduling and methods in Farneselli et al. [17]) using a synthetic fertilizer (Radicon N30, Green Has Italia spa, Italy) in the INT system and an organic fertilizer (Ilsadrip Forte, Ilsa spa, Italy) in the ORG and ORG+ systems. In the case of ORG, N content in the legume component of the CC (pea) was measured prior to termination and the corresponding amount was subtracted from the aimed rate of 150 kg N ha^{-1}. This difference was distributed to the cash crop. In the case of ORG+, only 50% of N accumulation in pea was subtracted from the aimed rate, in order to account for the lack of CC biomass incorporation into the soil [18].

Concerning the other macronutrients, 150 kg ha^{-1} of P_2O_5 and K_2O were broadcast at cover crop sowing (in ORG and ORG+) and at final seedbed preparation in INT.

2.1.2. Durum Wheat

Durum wheat was grown as the sole crop in the INT and ORG (single rows 0.15 m apart); in ORG+, durum wheat was temporary intercropped (TIC, Guiducci et al. [19]) with faba bean (*Vicia faba* L. var. *minor* Beck. cv Scuro di Torrelama) in alternate rows, 0.45 m apart with faba bean sown in the middle of the wheat inter-row space. Sowing density was 400 kernels m^{-2} for wheat (in all systems) and 90 seeds m^{-2} for faba bean (in ORG+).

Concerning wheat N fertilization, in the INT system, 160 kg N ha^{-1} was applied as urea in two applications (half dose at tillering and half at shooting, following the regional recommendation for durum wheat N fertilization management). In ORG, 40 kg N ha^{-1} was broadcast just before seedbed preparation as poultry manure (N = 4%). In ORG+, at the beginning of wheat shooting (Table S1), faba bean plants were incorporated into the top soil (0.10 m depth) by split rotary hoeing. Thus, in ORG+, durum wheat N fertilization came entirely from the incorporated faba bean plants (relying on an expected amount of approximately 50 kg N ha^{-1} [19])

2.2. Plant Sampling

Each year, the aboveground biomass accumulation of cash crops was determined before harvest by sampling plants from two subplots with 1.2 m^2 area per plot. The harvested aboveground biomass was separated in residues (straw, non-marketable fruits and vegetative parts) and yield (grains and marketable fruits). The cover crop and faba bean biomass was determined just before the termination (pea and barley in the mixture were kept separated). Weed biomass was also determined at each sampling operation. Plant samples were oven dried at 80 °C to determine dry matter content, then ground to a fine powder and stored. A reduced-N concentration of Kjeldhal digests, prepared following the method proposed by Isaac and Johnson [20], was measured by using an automatic analyzer (FlowSys, Systea, Italy).

2.3. NO$_3$-N Leaching, N Balance and N Use Efficiency

Every year, two lysimeters consisting of porous, ceramic cups (32 mm external diameter by 95 mm length) were installed [21] in the core part of each plot at a depth of 0.9 m. The cups were installed just after sowing by drilling the soil vertically at a depth of approximately 1.0 m. The excavated topsoil and lower subsoil were kept separate. Before placing the porous cup, thick slurry prepared from the lower subsoil was poured into the hole. The repacked soil was then added and consolidated with care in order to avoid preferential water flow. The ceramic cups (SDEC, Tauxigny, France) were joined to a capillary tube, long enough to emerge from the soil surface and sealed at the end by an iron clamp. Samples of the soil solution at 0.9 m were taken using a portable vacuum pump, and then transferred to a storage pot. The NO$_3$-N concentration in the soil solution was determined by an ion-specific electrode meter (Cardy, Spectrum Technologies, Inc., Plainfield, IL, USA), calibrated at the beginning of each measurement and set by using the standard solutions provided with the testing kits [22]. According to the method proposed by Gabriel et al. [23], NO$_3$-N concentration data were recorded only when all soil lysimeters could provide drainage water, which occurred after rainfall events of adequate intensity.

A simplified model was adopted to estimate the drainage volumes (for further details see Tosti et al. [22]). As proposed by Gabriel et al. [23], the NO$_3$-N leached over the time intervals between soil solution samplings was calculated as the product of mean NO$_3$-N concentration in the soil solution and the daily drainage obtained for the sampling interval.

As reported by De Notaris et al. [16], N balance was calculated as the difference between N input and output. The input included: N in manure or mineral fertilizer (i.e., extra-farm N), atmospheric N deposition [24], N derived from atmosphere via symbiotic fixation (Ndfa) and N in seeds. The output consisted of N removed from the field (i.e., leaching losses and N in yield). The N surplus was generated by the combination of the input and output values. For each year, inputs and outputs were determined, for each system, as averages across crops.

Ndfa was considered equal to 90% of the total N accumulation in the pea and faba bean above-ground biomass, according to the findings reported by Antichi [25] and Saia et al. [26] for similar climatic conditions. N use efficiency at system scale was assessed by two indices: yield to N leaching loss ratio (Y/Nloss, kg kg^{-1} N) and yield to extra-farm-N input (Y/Nextra, kg kg^{-1} N).

2.4. Statistical Analysis

Data were analyzed by using the following linear mixed model, following the rules suggested by Onofri et al. [27]. In particular, the field (two levels), the year (four levels) and the system (three levels) were added as fixed effects with all their two- and three-way interactions. It should be noted that the 'year × field' interaction corresponds to the crop effect, as the crops are univocally identified by one specific field in one specific year. The blocks within fields and the plots within blocks within fields were added as random effects to account for blocking units and repeated measures. The 'field × year × system' interaction was always significant and the corresponding means were compared by using a generalized multiple comparison procedure with multiplicity adjustment [28]. All analyses were conducted by using the R statistical environment [29].

3. Results

3.1. Weather Conditions

During the 4-year experiment (September 2013–August 2017), the average annual temperature was 15.3 °C, and the cumulated yearly precipitation was 865 mm. In all years, most rainfall events were observed from September to April (wet period, on average 602 mm cumulated rainfalls), and during this period, there was a clear gradient going from 2013/14 (extremely wet, 822 mm) to 2014/15 (highly wet, 645 mm), to 2015/16 (dry, 522 mm) and 2016/17 (very dry, 420 mm).

Two extreme events were recorded in the first and last years: a severe hail was recorded on 12 June 2014 and a late frost event with temperature of −1.44 °C was recorded on 22 April 2017 (Figure 1).

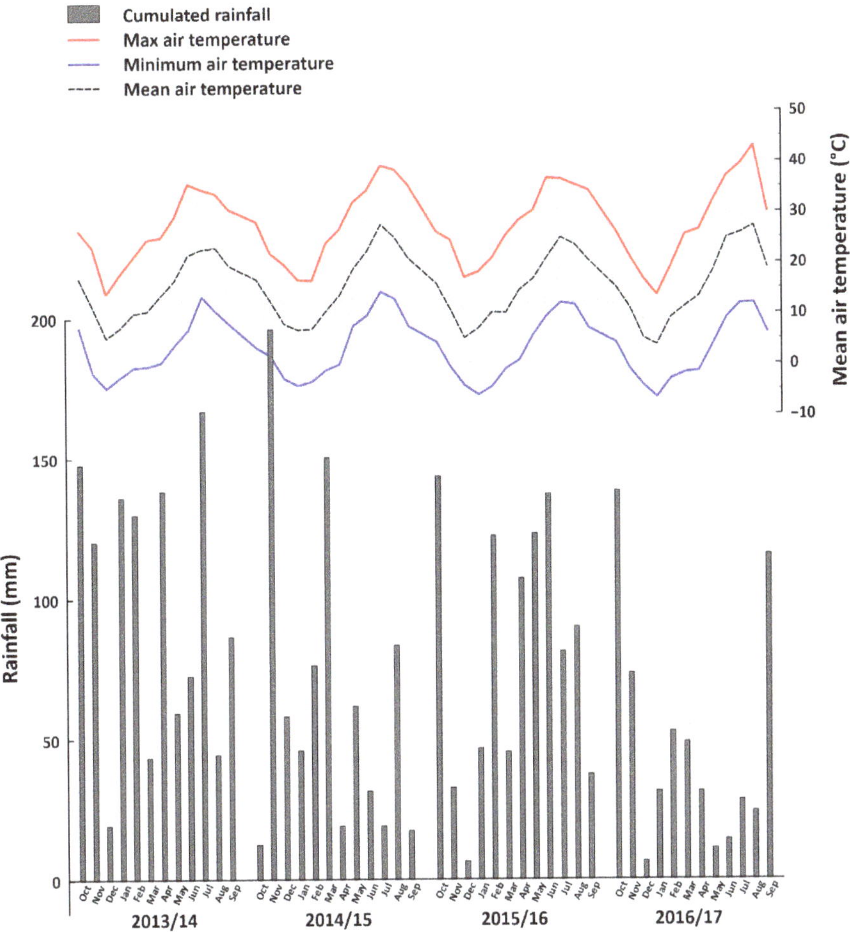

Figure 1. Monthly rainfall (mm) and mean, max and minimum air temperature values recorded at the experimental field station 'FieldLab' (Papiano, Perugia, Italy) during the 4-year experiment.

3.2. N Balance

In order to give an overview of the whole cropping system scale, N budgets were calculated for each cropping system as yearly averages on a 4-year basis (2013/14–2016/17, Table 1; the data for each main crop are reported in Table S2). The INT received a fixed amount of 155 kg N ha^{-1} yr^{-1} of extra-farm N fertilizer distributed to durum wheat (160 kg N ha^{-1} yr^{-1}) and processing tomato (150 kg N ha^{-1} yr^{-1}). The ORG received a fixed amount of N as poultry manure at durum wheat sowing, and a variable amount of extra-farm N via fertigation to the processing tomato (see Materials and methods section for details). Therefore, the yearly average of extra-farm N fertilizer added to ORG was 69.8 kg N ha^{-1} yr^{-1}. In ORG+, durum wheat was not fertilized with extra-farm source, however the N rates applied with fertigation to the processing tomato were generally higher than those in ORG; therefore, the yearly amount of N fertilizer added to ORG+ was, on average, 66.16 kg N ha^{-1} yr^{-1} (i.e., not statistically different from ORG, $p < 0.01$).

Table 1. Annual mean values (2-crop and 4-year basis, kg N ha^{-1} yr^{-1}) of Nitrogen inputs (Ninput), outputs (Noutput), surplus (Nsurplus) and N lost by leaching (Nleaching) during the experiment for the three cropping systems: integrated (INT), traditional organic (ORG) and innovative organic (ORG+). On the same row, values followed by different letters are statistically different ($p < 0.05$).

	INT		ORG		ORG+	
Ninput						
Fertilizers	155.0	a	69.8	b	66.2	b
Ndfa	0.0	-	25.2	b	34.2	a
Seeds	2.0	c	4.2	b	11.5	a
Deposition	5.0	a	5.0	a	5.0	a
Total Ninput	162.0	a	104.2	c	116.8	b
Noutput	104.0	a	75.7	b	44.0	c
Nsurplus	58.1	a	28.6	b	72.8	a
Nleaching	67.2	a	29.9	b	24.7	b

The highest Ndfa and N supplied with seeds were observed in ORG+ (legume component of CC + faba bean in TIC) followed by ORG. While the INT had no CC, neither legume was in crop rotation so the Ndfa was zero and the N added with the seeds was the lowest.

The INT was the system with the highest overall N input, while ORG had the lowest input. The average N output ranged from 44 kg N ha^{-1} yr^{-1} in ORG+ to 104 kg N ha^{-1} yr^{-1} in INT, while ORG showed an intermediate value of 76 kg N ha^{-1} yr^{-1}. INT and ORG+ had the highest N surplus values (58.1 and 72.8 kg N ha^{-1} yr^{-1}, respectively) while ORG the lowest (28.6 kg N ha^{-1} yr^{-1}). The N lost by leaching in INT was twofold compared to that observed in ORG and ORG+ (27.3 kg N ha^{-1} yr^{-1} on average, Table 1).

3.3. Yield and N Leaching

In all systems, durum wheat yield was low in 2013/14 and 2016/17 (due to a severe hail event and a late frost, respectively). In these two years, the systems did not show any significant difference in terms of yield (Figure 2A). In 2014/15 and 2015/16, the yields were generally higher and the ranking among systems was the same (i.e., INT > ORG > ORG+, $p < 0.05$).

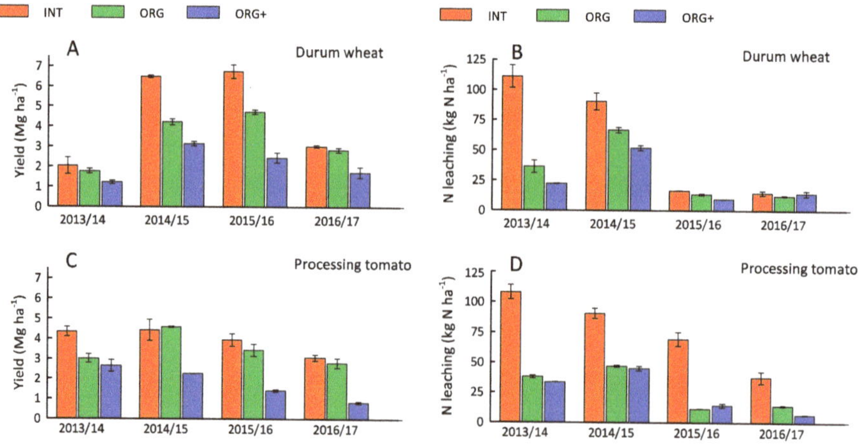

Figure 2. Yield (Mg ha^{-1}) and N leaching (kg N ha^{-1}) of durum wheat (top: (**A**) and (**B**), respectively) and processing tomato (bottom: (**C**) and (**D**), respectively) in the three cropping systems: integrated (INT), traditional organic (ORG) and innovative organic (ORG+). Bars represent the standard errors.

The N lost by leaching under durum wheat was significantly ($p < 0.001$) influenced by the systems in the first two (rainy) years, with the INT showing higher values as compared to both ORG and ORG+. In the latter two (drought) years, the N leaching values were generally low and the differences among systems were not significant (Figure 2B).

Processing tomato yield was compromised by the above-mentioned hail event in 2013/14, and the damages to the plants were particularly severe in ORG (Figure 2C). During the following years, yield values observed in INT and ORG were not different, while ORG+ always showed the lowest values. From 2014/15, the processing tomato yield showed a decreasing trend, particularly in ORG+, due to attacks of late blight disease (*Phytophthora infestans* (Mont.) de Bary) of increasing severity over time (Figure 2C).

In INT, where the soil was left bare during the autumn and winter seasons, N leaching in processing tomato decreased linearly across years as rainfall amount decreased (Figure 2D). Otherwise, irrespective of CC management strategies, both ORG and ORG+ showed similar N leaching values, which were significantly lower as compared to those observed in INT ($p < 0.001$).

3.4. N Loss and Extra-Farm N Input Per Yield Unit

In 2013/14 and 2014/15, the yield to N loss ratio (Y/Nloss, kg kg^{-1} N) of durum wheat was rather low (53 ± 3.8 kg kg^{-1} N) and it was not affected by the systems (Figure 3A). In 2015/16 and 2016/17, Y/Nloss was statistically similar in INT and ORG, which were significantly higher than ORG+. As for wheat, the Y/Nloss values of processing tomato (Figure 3C) observed in 2013/14 and 2014/15 were low in all systems: in 2013/14, the INT showed halved values as compared to ORG and ORG+, while in 2014/15 the systems did not show any significant difference (due to the high variability observed in INT). During 2015/16 and 2016/17, the Y/Nloss values observed in ORG were higher ($p < 0.001$) as compared to both the INT and ORG+, which were not significantly different from each other (Figure 3C).

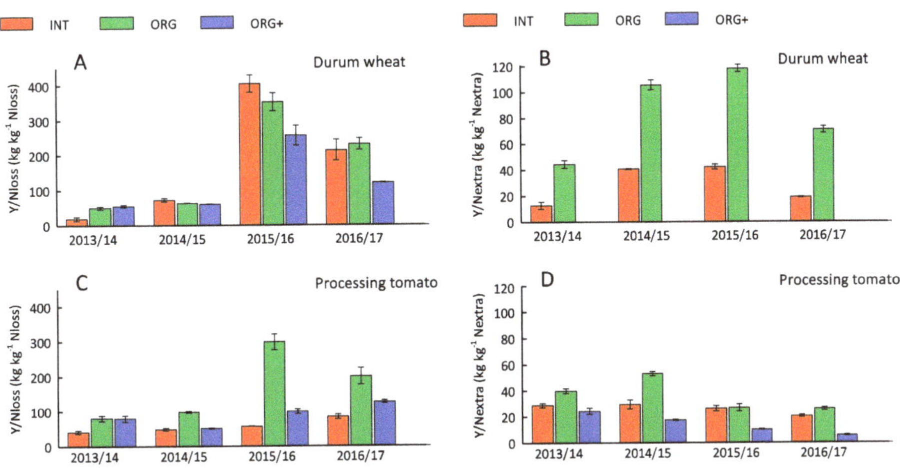

Figure 3. Yield per unit of N leached (Y/Nloss, kg kg^{-1} N), yield per unit of extra-farm N (Y/Nloss, kg kg^{-1} N) in durum wheat (top: (**A**) and (**B**), respectively) and processing tomato (bottom: (**C**) and (**D**), respectively) in the three cropping systems: integrated (INT), traditional organic (ORG) and innovative organic (ORG+). Bars represent the standard errors.

The yield to extra-farm–N input ratio (Y/Nextra, kg kg^{-1} N) of durum wheat was calculated just for INT and ORG systems, as the wheat in ORG+ did not receive any extra-farm N input (Figure 3B). Across the entire experimental period, the values observed in ORG were always significantly ($p < 0.001$) higher as compared to the INT. Concerning the Y/Nextra of processing tomato, in 2013/14 and 2014/15 the observed values in INT and ORG+ were not significantly different, but they were both lower than in ORG. In 2015/16 and 2016/17, ORG and INT were similar and higher than ORG+ (Figure 3D).

4. Discussion

The N balance (Table 1) allows the comparison between the three systems as it was computed by averaging the variations related to the conditions in individual years and crops [30]. The adoption of TIC in ORG+ considerably improved the N self-sufficiency of this system [31], raising (+35%) the Ndfa as compared to ORG. As expected, the Noutput values confirmed the gap between organic and conventional farming systems [32]; the Noutput values observed in ORG+ were the lowest in accordance with the finding of Knapp and van der Heijden [33] and Cooper et al. [34]. Nsurplus was high in both INT and ORG+, but such effects resulted from different reasons: in INT, there was a very high Ninput and a high Noutput, while in ORG+ there was high Ninput coupled with low Noutput. In contrast, Nsurplus in ORG was the lowest. The relation between N leaching and N surplus was consistent at the crop rotation level only in the INT, confirming that the management strategies to retain N in the system (e.g., by using catch crops and organic N fertilizers) are of paramount importance for reducing N leaching risk [16,35].

Wheat production in INT was the most variable across years, and such variability was associated with high N loss from the system when the climatic conditions were favorable to N leaching (high rainfall after side-dress fertilization and/or slow crop growth as recorded in the former two years, Figure 1). TIC was proved to reduce N leaching loss as compared to traditional management in organic wheat production [19,22]. Non-inversion soil management is known for reducing nutrient loss [36], thus its combination with TIC was expected to significantly improve the N retention in ORG+. However, such reduction (as compared to ORG) was not observed in our experiment, probably because the yield (and Noutput) was remarkably reduced in ORG+.

From 2014/15, a general reduction of processing tomato yield has been observed in all systems, probably due the short biannual rotation [37]. This effect was particularly evident in ORG+, where the short rotation problems were attributed to the typical decrease in crop productivity, during the transition from inversion to non-inversion soil till management [38]. In our case, the termination efficiency towards the CC, that is already known to be one of the critical issues in MBNT systems [8], was further reduced by the no-till soil management. However, processing tomato confirmed its good adaptability to organic practice [39] as the yields obtained in ORG were not statistically different from those observed in INT in 3 out of 4 years. As observed by other studies [32], when N-inputs are similar (such as for processing tomato), the yield gap between organic and conventional systems is lower (12% ± 5.0%) than when N-inputs differ (such as in durum wheat, 21% ± 4.8%). The introduction of winter-sown CC in crop rotation was confirmed to be a very effective practice to prevent N leaching [17,40]; the N loss in INT was proportionate to the rainfall amount during the rainy season (October to March, $R^2 = 0.854$, $n = 8$), while in ORG and ORG+ it was constantly low confirming the essential importance of CC for building agricultural systems with high N self-sufficiency and internal N recycling [41].

Recently, the yield gap in conventional and organic systems has been intensely debated [32,42]. In accordance with Wilbois and Schmidt [43], it is important to reframe this debate by taking into account the appropriate benchmarks. Thus, a comparison of the output in the three cropping systems cannot exclude the extra-farm–N input transformation efficiency (Y/NExtra) and the environmental cost (in terms of N lost from the system) per unit of yield (Y/NLoss).

In both wheat and processing tomato, ORG showed the best balance between economic output (yield) and water protection service as it showed the highest values of Y/NLoss in five cases out of eight (and in the three remaining cases the difference among systems was not significant). When favorable conditions for N leaching loss were present, the Y/Nloss observed in durum wheat was not affected by systems, while concerning processing tomato, the effect of the cover crop on N leaching reduction was predominant [44]. Therefore, although the INT showed the highest yield, the Y/Nloss ratio was not different from the very low yielding ORG+; on the other hand, ORG was the most interesting system in terms of environmental impact of the yield unit [45].

This interesting finding is confirmed by the Y/Nextra ratio (Figure 3B,D): considering durum wheat, the ORG efficiency in converting an extra-farm–N source to yield was largely above that showed by INT, while ORG+ was not considered, as it did not receive any external N input for wheat production. The efficiency of ORG+ was tested only in processing tomato, where it showed the lowest Y/Nextra values, due to the very low yield achieved by such system. Thus, at the cropping system level (i.e., considering both cash crops), this fact downplayed the impact of the complete N–self-sufficiency of ORG+ for durum wheat production. Thanks to its high yield, ORG showed Y/Nextra values that were not different (2013/14, 2015/16 and 2016/17) or even higher (2014/15) than those observed in INT.

5. Conclusions

Our results demonstrate that N recycling at the agroecosystem level was greatly improved by CC practice, independently from the soil management strategy (no till or inversion tillage). The improved Nsurplus was not retained in the system without CC (INT), so high yield (and Noutput) was not a sufficient condition to guarantee a high environmental efficiency of INT. On the contrary, the practices adopted in ORG+ (cover crop and temporary intercropping) considerably improved the N self-sufficiency of the system, thus the Nsurplus did not give rise to N loss, but the low Noutput (Yield) consistently reduced the efficiency of the external N input (Y/Nextra). The yield obtained per unit of N lost by leaching suggests that ORG was the most interesting system, while the potential sustainability of MBNT systems (i.e., ORG+) needs a substantial improvement, otherwise the great potential for the regulation and maintenance of ecosystem services can hardly be expressed.

Supplementary Materials: The following are available online at http://www.mdpi.com/2073-4395/9/11/718/s1, Table S1: Dates of the agronomic operations, Table S2: N balance with separate main crops.

Author Contributions: Conceptualization, G.T. and M.G.; data curation, G.T.; methodology, A.O.; software, A.O.; formal analysis, A.O.; writing—original draft preparation, G.T.; writing—review and editing, all authors; supervision, M.G.; project administration, G.T.; funding acquisition, G.T.

Funding: This research was carried out within the project SMOCA "Smart Management of Organic Conservation Agriculture" (http://smoca.agr.unipi.it/) funded by the Italian Ministry of University and Research (MIUR) within the program FIRB-2013 (Future in Research), MIUR-FIRB13 (project number: RBFR13L8J6).

Acknowledgments: We would like to thank: Patrizia Cannoni, Antonella Tizzi, Daniele Luchetti, Enzo Luchetti, Piero Bonciarelli, Fabio Ferretti and Daved Sforna for their invaluable help with the field work at the FieldLab in Papiano (PG, Italy).

Conflicts of Interest: The authors declare no conflict of interest.

References

1. Cammarano, D.; Ceccarelli, S.; Grando, S.; Romagosa, I.; Benbelkacem, A.; Akar, T.; Al-Yassin, A.; Pecchioni, N.; Francia, E.; Ronga, D. The impact of climate change on barley yield in the Mediterranean basin. *Eur. J. Agron.* **2019**, *106*, 1–11. [CrossRef]

2. Godfray, H.C.J.; Beddington, J.R.; Crute, I.R.; Haddad, L.; Lawrence, D.; Muir, J.F.; Pretty, J.; Robinson, S.; Thomas, S.M.; Toulmin, C. Food Security: The Challenge of Feeding 9 Billion People. *Science* **2010**, *327*, 812–818. [CrossRef] [PubMed]

3. Bommarco, R.; Kleijn, D.; Potts, S.G. Ecological intensification: Harnessing ecosystem services for food security. *Trends Ecol. Evol.* **2013**, *28*, 230–238. [CrossRef] [PubMed]

4. Duru, M.; Therond, O.; Martin, G.; Martin-Clouaire, R.; Magne, M.-A.; Justes, E.; Journet, E.-P.; Aubertot, J.-N.; Savary, S.; Bergez, J.-E.; et al. How to implement biodiversity-based agriculture to enhance ecosystem services: A review. *Agron. Sustain. Dev.* **2015**, *35*, 1259–1281. [CrossRef]

5. García-González, I.; Hontoria, C.; Gabriel, J.L.; Alonso-Ayuso, M.; Quemada, M. Cover crops to mitigate soil degradation and enhance soil functionality in irrigated land. *Geoderma* **2018**, *322*, 81–88. [CrossRef]

6. Benincasa, P.; Tosti, G.; Guiducci, M.; Farneselli, M.; Tei, F. Crop Rotation as a System Approach for Soil Fertility Management in Vegetables. In *Advances in Research on Fertilization Management of Vegetable Crops*; Advances in Olericulture; Springer: Cham, Switzerland, 2017; pp. 115–148. [CrossRef]

7. Carlesi, S.; Bigongiali, F.; Antichi, D.; Ciaccia, C.; Tittarelli, F.; Canali, S.; Bàrberi, P. Green manure and phosphorus fertilization affect weed community composition and crop/weed competition in organic maize. *Renew. Agr. Food Syst.* **2019**, 1–10. [CrossRef]

8. Vincent-Caboud, L.; Casagrande, M.; David, C.; Ryan, M.R.; Silva, E.M.; Peigne, J. Using mulch from cover crops to facilitate organic no-till soybean and maize production. A review. *Agron. Sustain. Dev.* **2019**, *39*, 45. [CrossRef]

9. Henneron, L.; Bernard, L.; Hedde, M.; Pelosi, C.; Villenave, C.; Chenu, C.; Bertrand, M.; Girardin, C.; Blanchart, E. Fourteen years of evidence for positive effects of conservation agriculture and organic farming on soil life. *Agron. Sustain. Dev.* **2015**, *35*, 169–181. [CrossRef]

10. Teasdale, J.R.; Coffman, C.B.; Mangum, R.W. Potential Long-Term Benefits of No-Tillage and Organic Cropping Systems for Grain Production and Soil Improvement. *Agron. J.* **2007**, *99*, 1297–1305. [CrossRef]

11. Peigné, J.; Vian, J.-F.; Payet, V.; Saby, N.P.A. Soil fertility after 10 years of conservation tillage in organic farming. *Soil Till. Res.* **2018**, *175*, 194–204. [CrossRef]

12. Bosco, S.; Volpi, I.; Antichi, D.; Ragaglini, G.; Frasconi, C. Greenhouse Gas Emissions from Soil Cultivated with Vegetables in Crop Rotation under Integrated, Organic and Organic Conservation Management in a Mediterranean Environment. *Agronomy* **2019**, *9*, 446. [CrossRef]

13. Casagrande, M.; Peigné, J.; Payet, V.; Mäder, P.; Sans, F.X.; Blanco-Moreno, J.M.; Antichi, D.; Bàrberi, P.; Beeckman, A.; Bigongiali, F.; et al. Organic farmers' motivations and challenges for adopting conservation agriculture in Europe. *Org. Agric.* **2016**, *6*, 281–295. [CrossRef]

14. Vincent-Caboud, L.; Peigné, J.; Casagrande, M.; Silva, E.M. Overview of Organic Cover Crop-Based No-Tillage Technique in Europe: Farmers' Practices and Research Challenges. *Agriculture* **2017**, *7*, 42. [CrossRef]

15. Kubota, H.; Iqbal, M.; Quideau, S.; Dyck, M.; Spaner, D. Agronomic and physiological aspects of nitrogen use efficiency in conventional and organic cereal-based production systems. *Renew. Agr. Food Syst.* **2018**, *33*, 443–466. [CrossRef]

16. De Notaris, C.; Rasmussen, J.; Sørensen, P.; Olesen, J.E. Nitrogen leaching: A crop rotation perspective on the effect of N surplus, field management and use of catch crops. *Agric. Ecosyst. Environ.* **2018**, *255*, 1–11. [CrossRef]

17. Farneselli, M.; Tosti, G.; Onofri, A.; Benincasa, P.; Guiducci, M.; Pannacci, E.; Tei, F. Effects of N sources and management strategies on crop growth, yield and potential N leaching in processing tomato. *Eur. J. Agron.* **2018**, *98*, 46–54. [CrossRef]

18. Montemurro, F.; Fiore, A.; Campanelli, G.; Tittarelli, F.; Ledda, L.; Canali, S. Organic Fertilization, Green Manure, and Vetch Mulch to Improve Organic Zucchini Yield and Quality. *HortScience* **2013**, *48*, 1027–1033. [CrossRef]

19. Guiducci, M.; Tosti, G.; Falcinelli, B.; Benincasa, P. Sustainable management of nitrogen nutrition in winter wheat through temporary intercropping with legumes. *Agron. Sustain. Dev.* **2018**, *38*, 31. [CrossRef]

20. Isaac, R.A.; Johnson, W.C. Determination of total nitrogen in plant tissue, using a block digestor. *J. Assoc. Off. Anal. Chem.* **1976**, *59*, 98–100.

21. Curley, E.M.; O'Flynn, M.G.; McDonnell, K.P. Porous Ceramic Cups: Preparation and Installation of Samplers for Measuring Nitrate Leaching. *Int. J. Soil Sci.* **2010**, *5*, 19–25. [CrossRef]

22. Tosti, G.; Farneselli, M.; Benincasa, P.; Guiducci, M. Nitrogen Fertilization Strategies for Organic Wheat Production: Crop Yield and Nitrate Leaching. *Agron. J.* **2016**, *108*, 1–12. [CrossRef]

23. Gabriel, J.L.; Munoz-Carpena, R.; Quemada, M. The role of cover crops in irrigated systems: Water balance, nitrate leaching and soil mineral nitrogen accumulation. *Agric. Ecosyst. Environ.* **2012**, *155*, 50–61. [CrossRef]

24. Ferretti, M.; Marchetto, A.; Arisci, S.; Bussotti, F.; Calderisi, M.; Carnicelli, S.; Cecchini, G.; Fabbio, G.; Bertini, G.; Matteucci, G.; et al. On the tracks of Nitrogen deposition effects on temperate forests at their southern European range—An observational study from Italy. *Glob. Chang. Biol.* **2014**, *20*, 3423–3438. [CrossRef] [PubMed]

25. Antichi, D. Effect of Intercropping on Yield and Quality of Organic Durum Wheat. Ph.D Thesis, University of Pisa, Pisa, Italy, 2013. Available online: http://etd.adm.unipi.it/ (accessed on 3 October 2019).

26. Saia, S.; Urso, V.; Amato, G.; Frenda, A.S.; Giambalvo, D.; Ruisi, P.; Miceli, G.D. Mediterranean forage legumes grown alone or in mixture with annual ryegrass: Biomass production, N2 fixation, and indices of intercrop efficiency. *Plant Soil* **2016**, *402*, 395–407. [CrossRef]

27. Onofri, A.; Seddaiu, G.; Piepho, H.-P. Long-Term Experiments with cropping systems: Case studies on data analysis. *Eur. J. Agron.* **2016**, *77*, 223–235. [CrossRef]

28. Bretz, F.; Hothorn, T.; Westfall, P.; Hothorn, T.; Westfall, P. *Multiple Comparisons Using R*; Chapman and Hall: New York, NY, USA, 2016. [CrossRef]

29. R Core Team. *R: A Language and Environment for Statistical Computing*; R Foundation for Statistical Computing: Vienna, Austria, 2018. Available online: https://www.R-project.org/ (accessed on 22 October 2019).

30. Jan, P.; Calabrese, C.; Lips, M. Determinants of nitrogen surplus at farm level in Swiss agriculture. *Nutr. Cycl. Agroecosyst.* **2017**, *109*, 133–148. [CrossRef]

31. Iannetta, P.P.M.; Young, M.; Bachinger, J.; Bergkvist, G.; Doltra, J.; Lopez-Bellido, R.J.; Monti, M.; Pappa, V.A.; Reckling, M.; Topp, C.F.E.; et al. A comparative nitrogen balance and productivity analysis of legume and non-legume supported cropping systems: The potential role of biological nitrogen fixation. *Front. Plant Sci.* **2016**, *7*. [CrossRef]

32. Schrama, M.; de Haan, J.J.; Kroonen, M.; Verstegen, H.; Van der Putten, W.H. Crop yield gap and stability in organic and conventional farming systems. *Agric. Ecosyst. Environ.* **2018**, *256*, 123–130. [CrossRef]

33. Knapp, S.; van der Heijden, M.G.A. A global meta-analysis of yield stability in organic and conservation agriculture. *Nat. Commun.* **2018**, *9*, 3632. [CrossRef]

34. Cooper, J.; Baranski, M.; Stewart, G.; Nobel-de Lange, M.; Bàrberi, P.; Fließbach, A.; Peigné, J.; Berner, A.; Brock, C.; Casagrande, M.; et al. Shallow non-inversion tillage in organic farming maintains crop yields and increases soil C stocks: A meta-analysis. *Agron. Sustain. Dev.* **2016**, *36*, 22. [CrossRef]

35. Reckling, M.; Hecker, J.-M.; Bergkvist, G.; Watson, C.A.; Zander, P.; Schläfke, N.; Stoddard, F.L.; Eory, V.; Topp, C.F.E.; Maire, J.; et al. A cropping system assessment framework—Evaluating effects of introducing legumes into crop rotations. *Eur. J. Agron.* **2016**. [CrossRef]

36. Blanco-Canqui, H.; Ruis, S.J. No-tillage and soil physical environment. *Geoderma* **2018**, *326*, 164–200. [CrossRef]

37. Marinari, S.; Mancinelli, R.; Brunetti, P.; Campiglia, E. Soil quality, microbial functions and tomato yield under cover crop mulching in the Mediterranean environment. *Soil Till. Res.* **2015**, *145*, 20–28. [CrossRef]

38. Alluvione, F.; Fiorentino, N.; Bertora, C.; Zavattaro, L.; Fagnano, M.; Chiaranda, F.Q.; Grignani, C. Short-term crop and soil response to C-friendly strategies in two contrasting environments. *Eur. J. Agron.* **2013**, *45*, 114–123. [CrossRef]

39. Farneselli, M.; Benincasa, P.; Tosti, G.; Pace, R.; Tei, F.; Guiducci, M. Nine-year results on maize and processing tomato cultivation in an organic and in a conventional low input cropping system. *Ital. J. Agron.* **2013**, *8*, e2. [CrossRef]

40. Tosti, G.; Benincasa, P.; Farneselli, M.; Tei, F.; Guiducci, M. Barley-hairy vetch mixture as cover crop for green manuring and the mitigation of N leaching risk. *Eur. J. Agron.* **2014**, *54*, 34–39. [CrossRef]

41. Drinkwater, L.E.; Snapp, S.S. Nutrients in Agroecosystems: Rethinking the Management Paradigm. In *Advances in Agronomy*; Sparks, D.L., Ed.; Academic Press: Cambridge, MA, USA, 2007; Volume 92, pp. 163–186. ISBN 9780080457420.

42. Connor, D.J. Organic agriculture cannot feed the world. *Field Crop. Res.* **2008**, *106*, 187–190. [CrossRef]

43. Wilbois, K.-P.; Schmidt, J.E. Reframing the Debate Surrounding the Yield Gap between Organic and Conventional Farming. *Agronomy* **2019**, *9*, 82. [CrossRef]

44. Wittwer, R.A.; Dorn, B.; Jossi, W.; van der Heijden, M.G.A. Cover crops support ecological intensification of arable cropping systems. *Sci. Rep.* **2017**, *7*, 41911. [CrossRef]

45. Dal Ferro, N.; Zanin, G.; Borin, M. Crop yield and energy use in organic and conventional farming: A case study in north-east Italy. *Eur. J. Agron.* **2017**, *86*, 37–47. [CrossRef]

Article

Agronomic Performances of Organic Field Vegetables Managed with Conservation Agriculture Techniques: A Study from Central Italy

Daniele Antichi [1,*], Massimo Sbrana [2], Luisa Martelloni [1], Lara Abou Chehade [1], Marco Fontanelli [1], Michele Raffaelli [1], Marco Mazzoncini [1], Andrea Peruzzi [1] and Christian Frasconi [1]

[1]   Department of Agriculture, Food and Environment, University of Pisa, via del Borghetto 80, 56124 Pisa, Italy; lmartelloni@agr.unipi.it (L.M.); lara.abouchehade@agr.unipi.it (L.A.C.); marco.fontanelli@unipi.it (M.F.); michele.raffaelli@unipi.it (M.R.); marco.mazzoncini@unipi.it (M.M.); andrea.peruzzi@unipi.it (A.P.); christian.frasconi@unipi.it (C.F.)

[2]   Center for Agro-Environmental Research "Enrico Avanzi", University of Pisa, via vecchia di Marina 6, 56122 San Piero a Grado, Pisa, Italy; massimosbrana75@gmail.com

*   Correspondence: daniele.antichi@unipi.it; Tel.: +39-050-221-8962

Received: 19 October 2019; Accepted: 25 November 2019; Published: 27 November 2019

Abstract: Organic farming systems are considered not compatible with conservation tillage mainly because of the reliance of conservative systems on herbicides. In this three-year field experiment, we tested the performances of an innovative vegetable organic and conservative system (ORG+) combining the use of cover crops (exploited as either living or dead mulch) and no-till techniques. This system was compared to "business-as-usual" organic farming (ORG) and integrated farming system (INT) based on the same crop sequence: savoy cabbage (*Brassica oleracea* var. sabauda L. cv. Famosa), spring lettuce (*Lactuca sativa* L. cv. Justine), fennel (*Foeniculum vulgare* Mill. Cv. Montebianco), and summer lettuce (*Lactuca sativa* L. cv. Ballerina RZ). The results of crop yield parameters and weed abundance contribute to spotlight potentialities and weaknesses of organic-conservative management of field vegetables. In particular, ORG+ caused significant yield depletion for all the crops and revealed suboptimal weed control and N availability. The agroecosystem services provided by the cover crops grown in the ORG+ as dead mulch or living mulch were affected by weather conditions and not always resulted in significant crop gain. Nevertheless, interesting results in terms of P availability and reduced N surplus encourage further development of the system targeting more sustainable organic vegetable production.

Keywords: sustainable agriculture; climate change mitigation; cabbage; fennel; lettuce; cover crops; green manure; no-till; dead mulch; living mulch; Mediterranean climate

---

1. Introduction

The response to a growing demand for food under climate change and environmental risks connected to intensive agriculture requires more efficient and sustainable agricultural practices. Ecological intensification has been proposed as a solution to these challenges, and organic farming principally relies on this, by promoting biodiversity and soil health [1–4]. Organic farming is increasingly growing in Europe, with almost 14.6 million hectares of agricultural land as of 2017 and a corresponding growth of more than 75% in a decade [5].

However, the current organic management practices have been an object of debate recently. Organic farmers still rely heavily on conventional intensive tillage to incorporate crop residues, organic fertilizers, and cover crops and most importantly to control weeds [6]. The intensive tillage leaves the soil exposed to wind and water erosion; destroys soil structure; and accelerates organic carbon loss due to oxidation, leaching, and translocation [7]. Reduced or no-tillage has been proposed in conventional agriculture to solve these problems. These practices are the key of "conservation agriculture" (CA), which not only contributes to sustain soil health and labor savings but also was shown to decrease greenhouse gas emissions through carbon sequestration and the reduction of energy use [8–10]. The implementation of reduced or no-tillage in organic agroecosystems may provide additional benefits to soil quality and may enhance resource use efficiency compared to conventional tillage systems [11,12]. However, these systems are challenged with nutrient availability at key crop growth stages and weed pressure, which are difficult to cope with in the absence of synthetic fertilizers, herbicides, and an adapted direct weed control [6]. These problems, often compromising crop yields, have limited so far the adoption of no-till by organic farmers, who in a recent survey showed interest to conservation practices for soil building purposes [13].

According to the Food and Agriculture Organization of the United Nations (FAO) definition [14], conservation agriculture systems imply also the diversification of cropping systems (e.g., by intercropping, cover cropping, agroforestry, and mixed farming) and permanent soil cover with crop residues or mulching material. The intensive use of cover crops has the potential to comply with these two principles. In organic farming, cover crops are normally used as green manures to provide nutrients to the cash crops and to increase soil fertility. Legume cover crops provide also by their N_2-fixing ability additional soil N for cash crops upon incorporation and decomposition [15–17]. Nevertheless, cover crops are well-known to provide a wide range of ecological services, such as protection against soil erosion, reduction of leaching and increased availability of nutrients, improvement of soil and water quality, and weed and pest control [18]. Moreover, it has been suggested that cover cropping effect would be higher by decreasing tillage intensity through an expected higher ecological intensification, which could alleviate weed and crop nutrition problems related to reduced tillage or no-till [19].

Cover crop-based no-till is one of the forms in which conservation practices can be integrated in organic farming. In these systems, cover crops could be grown to remain at the surface either as dead or living mulch when cash crop is to be planted [20]. Cover crops in no-till can reduce weed infestation during their growth and/or by their residues making a physical barrier, preventing sunlight reaching the soil surface or through allelopathy [21]. However, residues left on the soil surface generally slows down the decomposition rate and nitrogen release compared to their incorporation [22]. It has been shown the importance of cover crop management such as the selection, planting, and termination time and a complex rotation for the success of organic no-till, especially in vegetable systems which could suffer more from competition, nutrient shortage, and weeds [20].

On the other hand, the effects of conservation agriculture practices on soil greenhouse gas (GHG) emissions are still uncertain, especially when adopted within organic agriculture. Cover crops may impact soil processes in ways that could potentially increase or decrease GHG emissions [23]. Tillage also may indirectly affect GHG fluxes by altering soil biological and physical parameters in a variable way, demonstrating different responses across cropping systems [24]. Increased emissions in no-till farming were reported previously compared to conventionally tilled systems linked to soil types, climatic conditions, and the duration of conversion, although some studies showed lower emissions or no consistent effect [24–26].

Despite the environmental and economic promises that may hold, limited knowledge is available on organic cover crop-based no-till in Europe with far too little information concerning their performance in Mediterranean climate zone [18,26–28]. Thus, further investigation is needed as the success of organic reduced tillage systems may depend also on local pedoclimatic conditions. In this research, we aimed to study the transitional agronomic and environmental effects of the implementation of an organic conservation system (ORG+) within an organic vegetable rotation under Mediterranean conditions, with respect to an integrated management system (INT) and a standard organic one (ORG). Our objectives were to evaluate their performance in terms of (a) crop production, (b) nutrient availability, (c) N budget and use efficiency, and (d) weed infestation.

2. Materials and Methods

2.1. Site Characteristics

A three-year field experiment (2014–2017) was carried out on two adjacent fields (F1 and F2) at the Centre for Agri-environmental Research "Enrico Avanzi" of the University of Pisa, San Piero a Grado, Pisa, Italy (43°40′ N Lat; 10°19′ E Long; 1 m above mean sea level and 0% slope). The climate is typical Mediterranean with seasonal peaks of rainfall in spring and fall. Total average annual rainfall is 907 mm, and mean annual temperature is 15 °C. The soil was classified as Typic Xerofluvent, according to the USDA taxonomy [29]. The soil texture was loam sandy. Averaged over the two fields and two soil depths (i.e., 0–10 and 10–30 cm), soil organic matter content (Walkley–Black method) was 2.3 g 100 g^{-1} soil, total N (Kjeldahl method) was 1.25 mg g^{-1} soil, and available P (Olsen method) was 4.25 mg kg^{-1} soil. More details on the main parameters of soil fertility measured at the beginning of the experiment are reported in Reference [26].

2.2. Experimental Design and Crop Management

The field experiment was based on the following crop sequence: Savoy cabbage (*Brassica oleracea* var. sabauda L. cv. Famosa F1, Bejo), spring lettuce (*Lactuca sativa* L. cv. Justine, Clause), fennel (*Foeniculum vulgare* Mill. Cv. Montebianco F1, Olter), and summer lettuce (*L. sativa* L. cv. Ballerina RZ, Rijk Zwaan). The experimental field was split in two fields in order to rotate the crops both in space and time. For this reason, in 2014, the rotation started with fennel in the first field (F1) and with cabbage in the second field (F2). The experiment layout was explained in a recent paper focusing on GHG emissions in two years (2014–2016) [26]. Conversely, this paper covers the entire duration of the crop rotation that was replicated for three years (2014–2017). In F1, the crop sequence was fennel (2014), summer lettuce (2015), savoy cabbage (2015), spring lettuce (2016), fennel (2016), and summer lettuce (2017). In F2, the crop sequence was savoy cabbage (2014), spring lettuce (2015), fennel (2015), summer lettuce (2016), savoy cabbage (2016), and spring lettuce (2017).

Three different cropping systems were assigned to the experimental plots according to a randomized complete block (RCB) design [30] with one factor (i.e., the cropping system) and with three levels (i.e., the treatments) and three blocks per field following the main soil gradient of each field. The elementary plots were of size 63 m^2 (21 m long × 3 m wide) and were separated by alleyways of 3 m width within the blocks and 5 m width between the blocks.

The three cropping systems compared were a control, represented by an integrated farming system (INT) based on conventional tillage practices (i.e., spading, rotary cultivation, and hoeing), mechanical and chemical weed control, chemical pesticide, and mineral fertilizer (ammonium nitrate, superphosphate, and potassium sulphate) use; a standard organic cropping system (ORG), built upon the same tillage practices as INT, mechanical weed control, fertilization based on commercial solid organic fertilizers (pelleted dried manure, blood meal, rock phosphate, and potassium sulphate) and on the use of cover crops incorporated as green manures, and crop protection by substances and biocontrol agents admitted according to the Reg. CE 2007/834 and Reg. CE 2008/889; and an organic conservation system (ORG+) including continuous no-tillage, use of cover crops managed as living or dead mulches, reduced organic fertilizer application (same products used in ORG), cultural and thermal (i.e., flaming) weed control, and crop protection strategy as described for ORG.

The three tested cropping systems differed not only in terms of tillage intensity and use of agrochemicals but also more generally on external resource use. The ORG+ system was mainly designed to exploit internal natural resources. That is why fertilization levels were kept at a minimum, aiming to support naturally occurring soil-fertility-building processes rather than directly supplying nutrients to the crops. The total amounts of N supplied as fertilizers for the entire crop rotation cycle were 303, 155.6, and 55.5 kg N ha^{-1} respectively for INT, ORG, and ORG+. For P fertilizers, totals of 292, 192, and 87 kg P_2O_5 ha^{-1} were supplied respectively to INT, ORG, and ORG+. K fertilizers were supplied at 603, 385, and 120 kg K_2O ha^{-1} respectively to INT, ORG, and ORG+. The intensive use of cover crops in ORG+ was designed to replace several passes of mechanical weeding and herbicides. In the ORG system, a spring green manure mixture of field peas (*Pisum sativum* L.) and faba beans (*Vicia faba* var. *minor* Beck.) and a summer green manure mixture of red cowpea (*Vigna unguiculata* L. Walp), buckwheat (*Fagopyrum esculentum* L.), grain millet (*Panicum miliaceum* L.), and foxtail millet (*Setaria italica* L.) were chopped and incorporated into the soil before summer lettuce and fennel, respectively. In the ORG+ system, red clover (*Trifolium pratense* L.) was directly seeded and established as a living mulch for both summer lettuce and cabbage whilst the same summer cover crop mixture included in the ORG system was directly seeded on spring lettuce residues and terminated as dead mulch before the transplanting of fennel. In the ORG+ system, the termination of the dead mulch as well as the management of crop residues and living mulch was implemented by rolling with roller crimper alternated with flaming. The roller crimper used was the Clemens Eco-Roll (Clemens Technologies, Wittlich, Germany), whilst the flaming machine was a prototype developed by Maito (MAITO Srl., Arezzo, Italy). The two machines and operating conditions are described in detail in References [28,31] and in the Table S7. Direct transplanting of the field vegetables into the untilled soil in the ORG+ system was successfully implemented by using the modified version of the FAST transplanting machine produced by Fedele Mario Costruzioni (Fedele Costruzioni Meccaniche, Lanciano, Chieti, Italy) and developed by the University of Pisa [32].

Sprinkler irrigation was applied to all the treatments at the same volume in the ten days after transplant and, afterwards, every 3 days until harvest only during summer (May to September).

Detailed information on the management of each crop grown in each year/treatment in the two experimental fields is reported in Table S7.

2.3. Sampling Protocol and Measurements

Crop biomass production was assessed at harvest time (field vegetables) or before management (cover crops) by sampling the aboveground biomass of each crop on 3 areas per elementary plot. Plant samples collected in each sampling area were processed separately and the data were then averaged to obtain one value per each parameter per elementary plot (i.e., the block). The size of the sampling areas varied according to the spatial arrangement of the crops. For the cover crops, the biomass produced by each cover crop species and the total biomass were assessed on 0.5 m^2 (1 m wide × 0.5 m long) areas. For large row vegetables (i.e., fennel and savoy cabbage) transplanted at 0.75 m between the rows, the sampling areas covered two crop rows and were of the size 1.5 m^2 (1.5 m wide × 1 m long), including 4 and 8 plants per sampling area, respectively, for savoy cabbage and fennel. Likewise, for the lettuce crops, the sampling areas covered 2 rows but the size of the areas was 1 m^2 as the crops were transplanted with an inter-row space of 0.5 m. In this case, the total number of plants sampled was 10 per sampling area. In the same areas, the total aboveground biomass of weeds was also collected.

In the lab, crop and weed biomass were manually separated and the fresh weight of their total biomass was measured. Crop biomass was subdivided in marketable product (i.e., corymbs for savoy cabbage, swollen bases for fennel, and heads for lettuces) and residues (i.e., discarded products, outer leaves, and rotten/diseased/damaged plant biomass) that were fresh weighted separately.

The mean fresh weight of each marketable product unit (i.e., corymbs, swollen bases, and heads) was determined by dividing the total fresh weight of the marketable product by the number of product units. The mean dimension of the marketable products was assessed by measuring the lengths of the two main orthogonal diameters of each corymb, swollen basis, and head that were finally averaged to obtain one value for the mean diameter (cm) of the product unit of each crop.

From each of the three total plant samples collected in each plot, a representative subsample of each component (marketable product and residues, for the crops, and total aboveground biomass, for the weeds) was fresh weighted and oven-dried at 60 °C until constant weight. The dry material was then weighted to obtain the dry weight and the percentage of moisture. The Harvest Index was calculated as the ratio between the dry matter of marketable yield and the dry matter of total aboveground crop biomass. On the dry samples of the two components of the biomass of each vegetable crop (i.e., marketable product and residues), total nitrogen (Kjeldahl method) and total phosphorus (colorimetric method) concentration were determined.

2.4. Calculations

N and P_2O_5 accumulation in each biomass component were calculated as follows:

$$Nacc_i = Nconc_i \times dw_i \tag{1}$$

$$P_2O_5acc_i = Pconc_i \times dw_i \times 2.29 \tag{2}$$

where "$Nacc_i$" and "$P_2O_5acc_i$" are, respectively, the N and P_2O_5 accumulation (kg ha^{-1}) in the "i" biomass component of the crop (i.e., crop residues or marketable product); "$Nconc_i$" and "$Pconc_i$" are, respectively, the N and P concentration (g 100 g d.m^{-1}) in the "i" biomass component of the crop; and "dw_i" is the dry matter (kg ha^{-1}) in the "i" biomass component of the crop. The N and P_2O_5 accumulation in the total aboveground biomass of each crop was calculated as the sum of the accumulation in crop residues and marketable product.

The N budget (kg N ha^{-1}) at the level of single crops was estimated according to the following equation:

$$N_{budget} = (N_{fert} + N_{rain} + N_{min} + N_{fix}) - (N_{acc}) \tag{3}$$

where "Nfert" is the amount of N supplied by mineral and organic fertilizers (kg N ha^{-1}) applied to the single crop; "Nrain" is the amount of N supplied by rainfall that occurred in the growing period of the crop (kg N ha^{-1}), assuming that the mean N concentration in rain water is 3 mg N kg^{-1}; "Nmin" is the amount of N originated by the mineralization of the soil organic N in the first 30 cm of depth in the growing period of the crop (kg N ha^{-1}), assuming that the organic N content is 1.10 mg N kg^{-1} soil, the bulk density of the soil is 1.46 kg dm^{-3}, and the mineralization rate accounts for 2 g 100 g^{-1} year^{-1}; "Nfix" is the amount of N fixed from the atmosphere through symbiotic N_2 fixation of legume cover crops (i.e., red clover, pigeon bean, field pea, and red cowpea), assuming that the percentage of N derived from N_2 fixation on total N accumulated in the aboveground biomass of legumes is 80% and that there are no differences in the mineralization rate of legume cover crops managed as living/dead mulches or as green manures; and "Nacc" is the N accumulation in total biomass of the crops.

To assess the N use efficiency and to test whether N represented a limiting factor for crop yield in all the three systems, N surplus [33], N utilization efficiency (NUtE) [34], N Recovery Efficiency (NREac) [35], and Partial Factor Productivity (PFP) [36] were calculated as follows:

$$Nsurplus_{ti} = Nfert_i - Nacc_{ti} \tag{4}$$

$$Nsurplus_{yi} = Nfert_i - Nacc_{yi} \tag{5}$$

$$NUtE_i = Y_i/Nacc_{ti} \tag{6}$$

$$NREac_i = Nacc_{ti}/Ninput_i \tag{7}$$

$$NREac_{fi} = Nacc_{ti}/Nfert_i \tag{8}$$

$$PFP_i = Y_i/Ninput_i \tag{9}$$

$$PFP_{fi} = Y_i/Nfert_i \tag{10}$$

where "$Nfert_i$" is the N supplied as fertilizers (kg N ha^{-1}) to the "i" crop; "$Ninput_i$" is the total N input (kg N ha^{-1}) of the "i" crop; "$Nacc_{ti}$" is the N accumulated (kg N ha^{-1}) in total aboveground biomass of the "i" crop; "$Nacc_{yi}$" is the N accumulated (kg N ha^{-1}) in the marketable product of the "i" crop; and "Y_i" is the fresh weight of the marketable product of the "i" crop (Mg ha^{-1}).

2.5. Statistical Analysis

Data normality was assessed using the Shapiro–Wilk test. Other tests consisted of the Student's t-test to verify that the mean error was not significantly different to zero, the Breusch–Pagan test for homoscedasticity, and the Durbin–Watson test for autocorrelation.

All the dependent variables except for N use efficiency parameters were modelled in a linear mixed model using the extension package lmerTest (tests in the linear mixed effects models) [37] of R software [38]. We analyzed first the agronomic performances of each crop species (i.e., savoy cabbage, fennel, spring lettuce, and summer lettuce) separately in terms of fresh marketable yield (Y), dry matter of marketable yield (dw_y), dry matter of residues (dw_r), total aboveground dry matter (dw_t), harvest index (HI), total aboveground dry matter of weeds (dw_w), mean fresh weight of marketable product unit (MFW), mean diameter of marketable product unit (MD), N concentration in marketable yield ($Nconc_y$) and in residues ($Nconc_r$), N accumulation in marketable yield ($Nacc_y$), residues ($Nacc_r$) and total aboveground biomass ($Nacc_t$), P concentration in marketable yield ($Pconc_y$) and in residues ($Pconc_r$), P_2O_5 accumulation in marketable yield ($P_2O_5acc_y$), and residues ($P_2O_5acc_r$) and total aboveground biomass ($P_2O_5acc_t$). For these dependent variables, the cropping system and the year were the fixed factors whilst the block and the year were the random factors. The year was also tested as a fixed factor to test the effect of interannual variability on the dependent variables.

The agronomic performances of each cropping system at the level of entire crop sequence were analyzed separately for each field by summing the performances of each crop grown in the field over the entire experimental period (2014–2017). The global performances of the cropping systems were tested either including or not the contribution of cover crops in order to assess how they could lead to different performances in the systems. Cover crops affected only dry matter and nutrient parameters related to weed biomass (dw_w), crop residues (dw_r, $Nacc_r$), and total crop aboveground biomass (dw_t, $Nacc_t$). When analyzing these parameters as dependent variables, the cropping system, the inclusion/exclusion of cover crops and the field (i.e., F1 or F2) were the fixed factors and the block was the random factor. The field was considered as a factor as, in the crop sequence, there were slight differences in the number of occurrences of a single crop in the single field (Table S7). In the case of the analysis of parameters related to the crop marketable product (dw_y, $Nacc_y$, and $P_2O_5acc_y$), the cropping system and the field were the fixed factors and the block was the random factor. The effect of cover crops was not considered for these variables as they resulted from the sum of the marketable yield or the N accumulation in marketable yield of the vegetable crops and thus were not affected by the contribution of cover crops. Fitted correlations among the slopes were set. The analysis of variance was run.

3. Results

3.1. Weather Conditions

As shown in Figure 1, the weather conditions in the three experimental years differed from the normality for the area in many cases.

Monthly mean maximum temperatures were higher than multiannual values in most cases during the three experimental years. In particular, the winters were always warmer than usual, and so, it was also for the summer seasons in 2015 and 2017. Values below the multiannual means were registered in summer 2014 and in winter 2016/2017, instead. The hottest months in terms of maximum mean temperature were July 2015 and August 2017 with 32.4 °C. The lowest maximum mean temperature was registered in January 2017 (10.3 °C).

Likewise, the mean minimum temperatures were higher than multiannual values in the experimental period. Only in winter 2017, we observed values below the normality of the period. The coolest month was January 2017, indeed, with −0.8 °C, and the hottest was July 2016 (+20 °C).

The three experimental years were also characterized by high levels of rainfall compared to the multiannual trend. The rainiest months were January 2014 (355 mm), November 2014 (290 mm),

October 2015 (254 mm), and September 2017 (234 mm). Unusual high peaks of rainfall occurred also in July 2014, August 2015, June 2016, September 2016, and September 2017. The driest month was July 2017 when the experiment ended, with no rainfall registered at all.

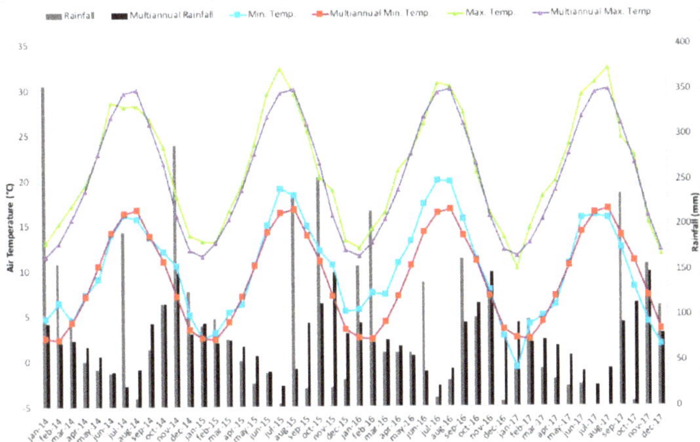

Figure 1. Monthly total rainfall (mm) and mean maximum and minimum air temperature (°C) from January 2014 to December 2017 compared to multiannual mean values (1993–2017).

3.2. Living Mulch, Dead Mulch, and Green Manure Biomass Production and NP Accumulations

The dry biomass produced by the spring cover crops never exceeded 3 Mg ha^{-1}, and it varied over years (Table 1). The mixture of field pea and pigeon bean grown in the ORG system clearly produced higher biomass than the red clover living mulch (ORG+), which was productive only in the first year. Red clover produced very low biomass especially in the second year, likely because of the low rainfall occurred in spring.

Table 1. Dry matter production of spring and summer cover crops and their weeds at termination dates in organic farming (ORG) and organic and conservative system (ORG+) in the three years. Means ± SE.

Year	Cropping Systems	Spring Cover Crop Aboveground Dry Matter (Mg ha^{-1}) [†]					Summer Cover Crop Aboveground Dry Matter (Mg ha^{-1}) [†]					
		Tp	*Vf*	*Ps*	Tot	We	*Vu*	*Fe*	*Pm*	*Si*	Tot	We
2014/15	ORG	-	1.26 ± 0.11	0.80 ± 0.15	2.06 ± 0.48	0.77 ± 0.06	0.75 ± 0.13	0.09 ± 0.03	1.12 ± 0.14	0.92 ± 0.11	2.88 ± 0.27	0.19 ± 0.05
	ORG+	2.27 ± 0.02	-	-	2.27 ± 0.02	2.59 ± 0.22	0.45 ± 0.06	0.14 ± 0.06	0.38 ± 0.11	0.34 ± 0.08	1.31 ± 0.37	3.16 ± 0.92
2015/16	ORG	-	1.81 ± 0.51	1.15 ± 0.21	2.96 ± 0.90	0.81 ± 0.25	1.10 ± 0.23	0.02 ± 0.01	0.44 ± 0.29	2.12 ± 0.36	3.68 ± 0.35	1.30 ± 0.43
	ORG+	0.09 ± 0.02	-	-	0.09 ± 0.02	n.a.[1]	0.25 ± 0.06	0.01 ± 0.00	0.17 ± 0.14	3.73 ± 0.82	4.16 ± 1.41	2.09 ± 0.50
2016/17	ORG	-	0.65 ± 0.08	0.51 ± 0.10	1.16 ± 0.10	0.74 ± 0.18	1.02 ± 0.28	0.00 ± 0.00	1.94 ± 0.22	1.77 ± 0.24	4.73 ± 0.11	0.47 ± 0.02
	ORG+	0.77 ± 0.28	-	-	0.77 ± 0.28	n.a.[1]	1.66 ± 0.33	0.00 ± 0.00	0.76 ± 0.13	1.13 ± 0.10	3.55 ± 0.06	1.32 ± 0.42

[1] n.a. is not available; [†] Tp is *Trifolium pratense*, Vf is *Vicia faba* var. minor, Ps is *Pisum sativum*, Vu is *Vigna unguiculata*, Fe is *Fagopyrum esculentum*, Pm is *Panicum miliaceum*, Si is *Setaria italica*, Tot is total biomass of cover crop mixtures, and We is total weed biomass.

The summer cover crop mixture was more productive and stable than the spring one. This was mainly because of the constantly high biomass production of foxtail millet and grain millet. Buckwheat biomass was very scarce, especially in the last two years, whereas red cowpea was steadily present over years at around 1 Mg ha^{-1} (Table 1).

3.3. Field Vegetable Biomass Production and NP Accumulations

The results of the statistical analysis of savoy cabbage, fennel, spring lettuce, and summer lettuce yield, biomass production, produce dimension, and NP concentration and accumulation are reported, respectively, in Tables S1–S4.

3.3.1. Savoy Cabbage

The cropping system had significant effects on all the tested variables except for the dry biomass of the weeds and the P concentration in the marketable product (Table S1). The year effect was not significant for fresh marketable yield, total dry matter, mean fresh weight of corymbs, the dry biomass of the weeds, and the accumulations of N and P_2O_5 in marketable yield and total biomass.

In Tables 2 and 3, the within-year effect of the cropping system for all the tested variables on crop biomass production is shown.

Table 2. Least squares means and standard errors of marketable fresh yield (Y), dry matter of marketable yield (dw$_y$), dry matter of residues (dw$_r$), total aboveground dry matter (dw$_t$), mean fresh weight of corymbs (MFW), Harvest Index (HI), and mean diameter of corymbs (MD) in savoy cabbage. Confidence level: 95%.

Dependent Variable	Lsmeans (±SE)		
	INT	**ORG**	**ORG+**
Y 2014 (Mg ha^{-1})	25.33 (2.80) a	25.39 (2.80) a	12.15 (2.80) b
Y 2015 (Mg ha^{-1})	22.53 (2.80) b	33.25 (2.80) a	14.62 (2.80) b
Y 2016 (Mg ha^{-1})	24.26 (2.80) a	24.98 (2.80) a	2.61 (3.43) * b
dw$_y$ 2014 (Mg ha^{-1})	1.91 (0.19) a	1.91 (0.19) a	0.91 (0.19) b
dw$_y$ 2015 (Mg ha^{-1})	1.76 (0.19) b	2.63 (0.19) a	1.64 (0.19) b
dw$_y$ 2016 (Mg ha^{-1})	1.92 (0.19) a	2.39 (0.19) a	0.27 (0.24) * b
dw$_r$ 2014 (Mg ha^{-1})	3.35 (0.26) a	3.76 (0.26) a	1.32 (0.26) b
dw$_r$ 2015 (Mg ha^{-1})	2.23 (0.26) a	2.37 (0.26) a	1.43 (0.26) b
dw$_r$ 2016 (Mg ha^{-1})	3.18 (0.26) a	2.75 (0.26) a	0.78 (0.26) b
dw$_t$ 2014 (Mg ha^{-1})	5.26 (0.43) a	5.67 (0.43) a	2.24 (0.43) b
dw$_t$ 2015 (Mg ha^{-1})	3.99 (0.43) ab	5.00 (0.43) a	3.08 (0.43) b
dw$_t$ 2016 (Mg ha^{-1})	5.10 (0.43) a	5.14 (0.43) a	0.96 (0.43) b
MFW 2014 (g)	949.99 (104.82) a	952.11 (104.82) a	455.74 (104.82) b
MFW 2015 (g)	844.99 (104.82) b	1246.65 (104.82) a	548.33 (104.82) b
MFW 2016 (g)	909.86 (104.82) a	936.81 (104.82) a	255.32 (128.37) b
HI 2014	0.36 (0.01) a	0.34 (0.01) a	0.42 (0.01) b
HI 2015	0.43 (0.01) a	0.52 (0.01) b	0.53 (0.01) b
HI 2016	0.37 (0.01) a	0.46 (0.01) b	0.26 (0.02) c
MD 2014 (cm)	17.42 (0.86) a	16.74 (0.86) a	11.25 (0.86) b
MD 2015 (cm)	19.90 (0.86) a	20.83 (0.86) a	16.42 (0.86) b
MD 2016 (cm)	19.21 (0.86) a	18.74 (0.86) a	12.88 (1.05) b

Means followed by different letters are statistically different (95% confidence interval). * Value statistically not different from zero.

In terms of fresh marketable product, the organic conservative system (ORG+) yielded significantly less than the standard organic (ORG) and the integrated system (INT) in all three years. The worst results were achieved in 2016, when the ORG+ yielded only 2.61 Mg ha^{-1} (a value that was even statistically not different from zero) as many cabbage plants in intercropping with the living mulch of red clover were not able to reach the corymb set stage. Averaged over the first two years, ORG+ yielded around 50% less than ORG and INT. Nevertheless, in 2015, ORG+ yielded statistically not different from INT both in terms of fresh (Y) and dry matter (dw_y) of marketable yield. In 2015, ORG significantly outyielded INT, whilst in the other two years, the yields of the two systems were comparable.

The dry matter of residues was always lower in ORG+ than in ORG and INT. As a result, the total aboveground dry matter produced by savoy cabbage in the three years followed the same trend, with lowest values in ORG+. Only in 2015, we did not observe any statistical differences between ORG+ and INT, with INT performing not differently from ORG. Interestingly, the HI revealed a substantial similarity between ORG and ORG+, with the two systems showing values significantly higher than INT in two years over three (i.e., in 2015 and 2016 for ORG and in 2014 and 2015 for ORG+). Only in 2016, due to the very low yield, ORG+ was significantly lower in terms of HI with respect to ORG and INT.

The mean fresh weight of corymbs followed the same trend as the biomass of marketable yield, with ORG+ showing significantly lower values than ORG (in all three years) and INT (in 2014 and 2016). The MFW in ORG+ was clearly lower if compared to ORG and INT.

The mean diameter of corymbs had a similar trend in all the three years, with ORG+ showing on average 5 cm lower values with respect to INT and ORG.

In Table 3, the within-year main effects of the cropping system on N and P concentration and accumulation in savoy cabbage biomass are shown.

Table 3. Least squares means and standard errors of N concentration in marketable yield ($Nconc_y$) and residues ($Nconc_r$); N accumulation in marketable yield ($Nacc_y$), residues ($Nacc_r$), and total aboveground dry matter ($Nacc_t$); P concentration in marketable yield ($Pconc_y$) and residues ($Pconc_r$); and P_2O_5 accumulation in marketable yield ($P_2O_5acc_y$), residues ($P_2O_5acc_r$), and total aboveground dry matter ($P_2O_5acc_t$) in savoy cabbage. Confidence level: 95%.

Dependent Variable	Lsmeans (±SE)		
	INT	**ORG**	**ORG+**
$Nconc_y$ (g 100 g^{-1}) 2014	2.68 (0.11) a	2.72 (0.11) a	4.04 (0.11) b
$Nconc_y$ (g 100 g^{-1}) 2015	3.31 (0.11) a	2.37 (0.11) b	1.79 (0.11) c
$Nconc_y$ (g 100 g^{-1}) 2016	2.97 (0.11) a	2.54 (0.11) b	3.14 (0.14) a
$Nconc_r$ (g 100 g^{-1}) 2014	2.63 (0.09) a	1.85 (0.09) b	2.81 (0.09) a
$Nconc_r$ (g 100 g^{-1}) 2015	2.91 (0.09) a	2.12 (0.09) b	1.56 (0.09) c
$Nconc_r$ (g 100 g^{-1}) 2016	2.55 (0.09) a	2.05 (0.09) b	2.44 (0.09) a
$Nacc_y$ (kg ha^{-1}) 2014	51.62 (6.19) a	51.91 (6.19) a	36.22 (6.19) a
$Nacc_y$ (kg ha^{-1}) 2015	58.31 (6.19) a	61.92 (6.19) a	29.41 (6.19) b
$Nacc_y$ (kg ha^{-1}) 2016	57.19 (6.19) a	60.38 (6.19) a	8.40 (7.58) b
$Nacc_r$ (kg ha^{-1}) 2014	88.43 (6.98) a	69.84 (6.98) a	37.84 (6.98) b
$Nacc_r$ (kg ha^{-1}) 2015	64.53 (6.98) a	50.27 (6.98) a	22.31 (6.98) b
$Nacc_r$ (kg ha^{-1}) 2016	81.48 (6.98) a	56.79 (6.98) a	19.20 (6.98) b
$Nacc_t$ (kg ha^{-1}) 2014	140.04 (12.50) a	121.75 (12.50) a	74.06 (12.50) b
$Nacc_t$ (kg ha^{-1}) 2015	122.83 (12.50) a	112.19 (12.50) a	51.73 (12.50) b
$Nacc_t$ (kg ha^{-1}) 2016	138.67 (12.50) a	117.17 (12.50) a	24.80 (12.50) * b
$Pconc_y$ (g 100 g^{-1}) 2014	0.34 (0.01) a	0.32 (0.01) a	0.35 (0.01) b
$Pconc_y$ (g 100 g^{-1}) 2015	0.31 (0.01) a	0.33 (0.01) b	0.30 (0.01) a
$Pconc_y$ (g 100 g^{-1}) 2016	0.32 (0.01) a	0.33 (0.01) a	0.32 (0.01) a
$Pconc_r$ (g 100 g^{-1}) 2014	0.23 (0.01) a	0.22 (0.01) a	0.28 (0.01) b
$Pconc_r$ (g 100 g^{-1}) 2015	0.19 (0.01) a	0.21 (0.01) a	0.19 (0.01) a
$Pconc_r$ (g 100 g^{-1}) 2016	0.21 (0.01) a	0.22 (0.01) a	0.25 (0.01) b
$P_2O_5acc_y$ (kg ha^{-1}) 2014	14.87 (1.52) a	14.27 (1.52) a	7.36 (1.52) b
$P_2O_5acc_y$ (kg ha^{-1}) 2015	12.45 (1.52) a	19.92 (1.52) b	11.22 (1.52) a
$P_2O_5acc_y$ (kg ha^{-1}) 2016	14.18 (1.52) a	18.10 (1.52) a	2.01 (1.87) * b

Table 3. *Cont.*

Dependent Variable	Lsmeans (±SE)		
	INT	**ORG**	**ORG+**
$P_2O_5acc_r$ (kg ha^{-1})2014	17.96 (1.73) a	19.46 (1.73) a	8.78 (1.73) b
$P_2O_5acc_r$ (kg ha^{-1})2015	9.66 (1.73) ab	11.42 (1.73) a	6.29 (1.73) b
$P_2O_5acc_r$ (kg ha^{-1})2016	15.53 (1.73) a	13.92 (1.73) a	4.47 (1.73) b
$P_2O_5acc_t$ (kg ha^{-1})2014	32.83 (3.04) a	33.73 (3.04) a	16.14 (3.04) b
$P_2O_5acc_t$ (kg ha^{-1})2015	22.11 (3.04) a	31.35 (3.04) b	17.51 (3.04) a
$P_2O_5acc_t$ (kg ha^{-1})2016	29.71 (3.04) a	32.02 (3.04) a	5.80 (3.04) * b

Means followed by different letters are statistically different (95% confidence interval). * Value statistically not different from zero.

N concentration in corymbs and residues was normally higher in INT than ORG, except for 2014, when we did not observe any difference for marketable yield. The cabbage grown in ORG+ plots showed values of N concentration lower than ORG and INT only in 2015, whilst in 2014, it showed the highest value for $Nconc_y$ and, in 2016, it performed equal to the other two systems. As a result of the combination between concentration and dry matter production, N accumulation showed overall significantly lower values in ORG+ than ORG and INT. Only for $Nacc_y$ in 2014, we observed comparable results among the three systems.

For P concentration, ORG+ did not show lower values compared to ORG and INT. In 2014, both $Pconc_y$ and $Pconc_r$ were higher in ORG+ than ORG and INT. Also, in 2015, $Pconc_r$ was higher in ORG+. P accumulation in the dry matter of savoy cabbage was affected by dry matter production values and revealed normally lower values in ORG+ than ORG and INT but with some exceptions in 2015, when $P_2O_5acc_y$ and $P_2O_5acc_t$ and $P_2O_5acc_r$ were not different in ORG+ and INT. Oppositely to N accumulation, in absolute terms, ORG produced slightly higher values of P accumulation in savoy cabbage total biomass with respect to INT and ORG+.

3.3.2. Fennel

For fennel, the effect of the cropping system was significant for all the dependent variables tested except for HI and $P_2O_5acc_r$ (Table S2). The year effect was not significant only for $Nacc_r$, $Pconc_y$, $Pconc_r$, $P_2O_5acc_r$, and $P_2O_5acc_t$.

The data on biomass production of fennel at harvest time in all three years are reported in Table 4. For this crop, there were no significant differences between INT and ORG, although the integrated system always resulted in the highest values of fresh and dry marketable yield. For the organic conservative system, we observed encouraging results, as the fennel in ORG+ plots performed statistically equal to ORG in 2014 and 2015, although always significantly lower than INT. In 2015, the fresh dry matter of swollen bases collected in the ORG+ plots was even higher than that in the ORG plots, although not significantly. The same trend was observed also for the mean fresh weight of the swollen bases.

Overall, the yield depletion observed in the ORG+ system, compared to ORG and INT, averaged ca. 35%. The dry matter of residues in ORG+ was statistically not different from ORG only in 2015 and was always lower than INT. The dry matter of the total aboveground biomass was not different between INT and ORG, whereas it was significantly lower in ORG+. For HI, as mentioned, we did not observe any differences among the three cropping systems. The mean diameter of the swollen bases was significantly affected by the cropping systems and the highest values were always shown by INT. The ORG system produced bases with similar diameters to INT in 2014 and 2016. In 2015, ORG was significantly lower than ORG+ and INT, instead.

Data on N and P concentration and accumulation in biomass components of fennel under the three cropping systems are shown in Table 5. In 2014 and 2016, N concentration was higher in INT than ORG heads and in ORG than ORG+ heads. In 2015, ORG+ showed higher values than ORG. For N concentration in crop residues, the INT system showed significantly higher values than ORG and

ORG+ in 2015 and 2016. In 2014, there were no differences between INT and ORG, only with ORG+. In 2015, ORG and ORG+ showed similar results. N accumulation in marketable dry matter was higher in INT than ORG in all three years. Only in 2015, we did not observe any differences between ORG and ORG+. The same trend was observed also for N accumulation in crop residues, a parameter for which there were no significant differences between INT and ORG in 2014 and 2016. For P uptake, the differences among the three systems were less evident. ORG+ showed similar results to INT in many cases (i.e., for $Pconc_y$ in all three years, for $Pconc_r$ in 2015, for $P_2O_5acc_y$ in 2014, and for $P_2O_5acc_r$ and $P_2O_5acc_t$ in all three years).

Table 4. Least squares means and standard errors of marketable fresh yield (Y), dry matter of marketable yield (dw_y), dry matter of residues (dw_r), total aboveground dry matter (dw_t), mean fresh weight of swollen bases (MFW), Harvest Index (HI), and mean diameter of swollen bases (MD) in fennel. Confidence level: 95%.

Dependent Variable	Lsmeans (±SE)		
	INT	**ORG**	**ORG+**
Y 2014 (Mg ha^{-1})	12.33 (1.27) a	10.08 (1.27) ab	6.96 (1.27) b
Y 2015 (Mg ha^{-1})	21.68 (1.27) a	13.77 (1.27) b	14.82 (1.27) b
Y 2016 (Mg ha^{-1})	16.03 (1.27) a	12.55 (1.27) a	6.62 (1.27) b
dw_y 2014 (Mg ha^{-1})	0.64 (0.06) a	0.50 (0.06) ab	0.44 (0.06) b
dw_y 2015 (Mg ha^{-1})	0.91 (0.06) a	0.88 (0.06) a	0.65 (0.06) b
dw_y 2016 (Mg ha^{-1})	0.77 (0.06) a	0.76 (0.06) a	0.38 (0.06) b
dw_r 2014 (Mg ha^{-1})	1.20 (0.10) a	1.21 (0.10) a	0.76 (0.10) b
dw_r 2015 (Mg ha^{-1})	1.55 (0.10) a	1.31 (0.10) ab	1.10 (0.10) b
dw_r 2016 (Mg ha^{-1})	1.55 (0.10) a	1.52 (0.10) a	0.76 (0.10) b
dw_t 2014 (Mg ha^{-1})	1.84 (0.14) a	1.71 (0.14) a	1.20 (0.14) b
dw_t 2015 (Mg ha^{-1})	2.47 (0.14) a	2.18 (0.14) a	1.75 (0.14) b
dw_t 2016 (Mg ha^{-1})	2.32 (0.14) a	2.28 (0.14) a	1.14 (0.14) b
MFW 2014 (g)	231.11 (23.85) a	188.93 (23.85) ab	130.58 (23.85) b
MFW 2015 (g)	406.53 (23.85) a	258.12 (23.85) b	277.85 (23.85) b
MFW 2016 (g)	300.56 (23.85) a	235.28 (23.85) a	124.03 (23.85) b
HI 2014	0.35 (0.02) a	0.30 (0.02) b	0.37 (0.02) a
HI 2015	0.37 (0.02) a	0.40 (0.02) a	0.37 (0.02) a
HI 2016	0.33 (0.02) a	0.33 (0.02) a	0.33 (0.02) a
MD 2014 (cm)	6.68 (0.36) a	6.62 (0.36) a	4.24 (0.36) b
MD 2015 (cm)	9.58 (0.36) a	8.00 (0.36) b	8.58 (0.36) a
MD 2016 (cm)	7.86 (0.36) a	6.84 (0.36) a	4.53 (0.36) b

Means followed by different letters are statistically different (95% confidence interval).

Table 5. Least squares means and standard errors of N concentration in marketable yield ($Nconc_y$) and residues ($Nconc_r$); N accumulation in marketable yield ($Nacc_y$), residues ($Nacc_r$), and total aboveground dry matter ($Nacc_t$); P concentration in marketable yield ($Pconc_y$) and residues ($Pconc_r$); and P_2O_5 accumulation in marketable yield ($P_2O_5acc_y$), residues ($P_2O_5acc_r$), and total aboveground dry matter ($P_2O_5acc_t$) in fennel. Confidence level: 95%.

Dependent Variable	Lsmeans (±SE)		
	INT	**ORG**	**ORG+**
$Nconc_y$ (g 100 g^{-1}) 2014	2.88 (0.10) a	2.47 (0.10) b	1.51 (0.10) c
$Nconc_y$ (g 100 g^{-1}) 2015	1.93 (0.10) a	1.17 (0.10) b	1.57 (0.10) c
$Nconc_y$ (g 100 g^{-1}) 2016	2.03 (0.10) a	1.42 (0.10) b	1.09 (0.10) c
$Nconc_r$ (g 100 g^{-1}) 2014	2.79 (0.14) a	3.06 (0.14) a	2.20 (0.14) b
$Nconc_r$ (g 100 g^{-1}) 2015	2.26 (0.14) a	1.61 (0.14) b	1.63 (0.14) b
$Nconc_r$ (g 100 g^{-1}) 2016	2.92 (0.14) a	2.37 (0.14) b	1.68 (0.14) c

Table 5. *Cont.*

Dependent Variable	Lsmeans (±SE)		
	INT	**ORG**	**ORG+**
$Nacc_y$ (kg ha^{-1}) 2014	18.53 (1.37) a	12.40 (1.37) b	6.55 (1.37) c
$Nacc_y$ (kg ha^{-1})2015	17.72 (1.37) a	10.38 (1.37) b	10.35 (1.37) b
$Nacc_y$ (kg ha^{-1})2016	15.42 (1.37) a	10.70 (1.37) b	4.13 (1.37) c
$Nacc_r$ (kg ha^{-1})2014	34.43 (3.38) a	37.69 (3.38) a	16.70 (3.38) b
$Nacc_r$ (kg ha^{-1})2015	35.27 (3.38) a	21.03 (3.38) b	17.87 (3.38) b
$Nacc_r$ (kg ha^{-1})2016	45.36 (3.38) a	35.76 (3.38) a	12.72 (3.38) b
$Nacc_t$ (kg ha^{-1})2014	52.97 (4.08) a	50.10 (4.08) a	23.26 (4.08) b
$Nacc_t$ (kg ha^{-1})2015	52.99 (4.08) a	31.41 (4.08) b	28.22 (4.08) b
$Nacc_t$ (kg ha^{-1})2016	60.78 (4.08) a	46.47 (4.08) b	16.84 (4.08) c
$Pconc_y$ (g 100 g^{-1}) 2014	0.41 (0.03) a	0.48 (0.03) a	0.44 (0.03) a
$Pconc_y$ (g 100 g^{-1}) 2015	0.50 (0.03) a	0.37 (0.03) b	0.55 (0.03) a
$Pconc_y$ (g 100 g^{-1}) 2016	0.45 (0.03) a	0.35 (0.03) b	0.49 (0.03) a
$Pconc_r$ (g 100 g^{-1}) 2014	0.34 (0.04) a	0.30 (0.04) a	0.41 (0.04) b
$Pconc_r$ (g 100 g^{-1}) 2015	0.29 (0.04) a	0.24 (0.04) a	0.33 (0.04) a
$Pconc_r$ (g 100 g^{-1}) 2016	0.27 (0.04) a	0.24 (0.04) a	0.47 (0.04) b
$P_2O_5acc_y$ (kg ha^{-1})2014	5.82 (0.69) a	5.61 (0.69) a	4.40 (0.69) a
$P_2O_5acc_y$ (kg ha^{-1})2015	10.40 (0.69) a	7.49 (0.69) b	8.26 (0.69) b
$P_2O_5acc_y$ (kg ha^{-1})2016	7.81 (0.69) a	6.00 (0.69) a	4.25 (0.69) b
$P_2O_5acc_r$ (kg ha^{-1})2014	10.27 (1.42) a	8.41 (1.42) a	7.10 (1.42) a
$P_2O_5acc_r$ (kg ha^{-1})2015	10.30 (1.42) a	7.10 (1.42) a	8.41 (1.42) a
$P_2O_5acc_r$ (kg ha^{-1})2016	9.596 (1.42) a	8.33 (1.42) a	8.09 (1.42) a
$P_2O_5acc_t$ (kg ha^{-1})2014	16.09 (1.83) a	14.02 (1.83) a	11.50 (1.83) a
$P_2O_5acc_t$ (kg ha^{-1})2015	20.70 (1.83) a	14.60 (1.83) b	16.67 (1.83) ab
$P_2O_5acc_t$ (kg ha^{-1})2016	17.41 (1.83) a	14.33 (1.83) a	12.34 (1.83) a

Means followed by different letters are statistically different (95% confidence interval).

3.3.3. Spring Lettuce

For the spring lettuce crop grown before fennel, the effect of the cropping system was not significant only for the P concentration in marketable yield. Besides P concentration in heads, the year did not significantly affect also the harvest index (HI) and the N accumulation in heads ($Nacc_y$) (Table S3).

The results of biomass production of spring lettuce at harvest time are shown in Table 6. For this crop, the performances of the ORG+ system were particularly negative. All the biomass components were significantly depleted by the ORG+ system, which was always lower than ORG and INT.

The concentration of N in heads and residues was highest in INT in all three years. INT did not differ from ORG only in 2017 whilst differed from ORG+ in 2015 and in 2017 (for crop residues) (Table 7). The amount of N accumulated in marketable product and residues was significantly higher in INT and ORG than ORG+ in all three years. $Nacc_r$ and $Nacc_t$ were not statistically lower in ORG than INT only in 2017. For P, the concentration in heads of lettuce was higher in ORG and ORG+ than INT in 2015, lower in ORG and ORG+ than INT in 2016, and not different among the systems in 2017. For crop residues, the P concentration was significantly higher in ORG+ than ORG and INT. The P accumulation in marketable yield was higher in INT than ORG and in ORG than ORG+ in 2015 and 2016. In 2017, there were no differences between INT and ORG, which both outperformed ORG+. For crop residues, INT was still the treatment with the highest P accumulation levels, being higher than ORG and ORG+ in 2015 and 2016 and higher than ORG+ alone in 2017. ORG and ORG+ did not differ from each other only in 2015. As a result, the total accumulation of P in the aboveground biomass of lettuce was higher in INT and ORG than ORG+ in 2014 and 2016 and was not different between INT and ORG only in 2016. P accumulation in total biomass in ORG+ was always lower than ORG.

Table 6. Least squares means and standard errors of marketable fresh yield (Y), dry matter of marketable yield (dw_y), dry matter of residues (dw_r), total aboveground dry matter (dw_t), mean fresh weight of heads (MFW), Harvest Index (HI), and mean diameter of heads (MD) in spring lettuce. Confidence level: 95%.

Dependent Variable	Lsmeans ($\pm$SE)		
	INT	ORG	ORG+
Y 2015 (Mg ha^{-1})	22.11 (2.29) a	21.65 (2.29) a	8.41 (2.29) b
Y 2016 (Mg ha^{-1})	32.21 (2.29) a	23.72 (2.29) b	5.20 (2.29) c
Y 2017 (Mg ha^{-1})	16.83 (2.29) a	14.79 (2.29) a	7.86 (2.29) b
dw_y 2015 (Mg ha^{-1})	1.08 (0.10) a	1.08 (0.10) a	0.55 (0.10) b
dw_y 2016 (Mg ha^{-1})	1.63 (0.10) a	1.53 (0.10) a	0.38 (0.10) b
dw_y 2017 (Mg ha^{-1})	1.19 (0.10) a	1.10 (0.10) a	0.56 (0.10) b
dw_r 2015 (Mg ha^{-1})	0.59 (0.04) a	0.43 (0.04) b	0.24 (0.04) c
dw_r 2016 (Mg ha^{-1})	1.25 (0.04) a	0.78 (0.04) b	0.17 (0.04) c
dw_r 2017 (Mg ha^{-1})	0.69 (0.04) a	0.67 (0.04) a	0.23 (0.04) b
dw_t 2015 (Mg ha^{-1})	1.67 (0.11) a	1.51 (0.11) a	0.78 (0.11) b
dw_t 2016 (Mg ha^{-1})	2.89 (0.11) a	2.31 (0.11) b	0.54 (0.11) c
dw_t 2017 (Mg ha^{-1})	1.88 (0.11) a	1.77 (0.11) a	0.78 (0.11) b
MFW 2015 (g)	221.13 (22.85) a	216.54 (22.85) a	84.10 (22.85) b
MFW 2016 (g)	322.12 (22.85) a	237.23 (22.85) b	51.96 (22.85) c
MFW 2017 (g)	168.28 (22.85) a	147.84 (22.85) a	78.60 (22.85) b
HI 2015	0.63 (0.03) a	0.72 (0.03) ab	0.70 (0.03) a
HI 2016	0.56 (0.03) a	0.66 (0.03) b	0.71 (0.03) b
HI 2017	0.63 (0.03) ab	0.62 (0.03) a	0.71 (0.03) b
MD 2015 (cm)	15.21 (0.59) a	15.76 (0.59) a	11.63 (0.59) b
MD 2016 (cm)	17.87 (0.59) a	17.23 (0.59) a	9.95 (0.59) b
MD 2017 (cm)	7.81 (0.59) a	7.76 (0.59) a	4.37 (0.59) b

Means followed by different letters are statistically different (95% confidence interval).

Table 7. Least squares means and standard errors of N concentration in marketable yield ($Nconc_y$) and residues ($Nconc_r$); N accumulation in marketable yield ($Nacc_y$), residues ($Nacc_r$), and total aboveground dry matter ($Nacc_t$); P concentration in marketable yield ($Pconc_y$) and residues ($Pconc_r$); P_2O_5 accumulation in marketable yield ($P_2O_5acc_y$), residues ($P_2O_5acc_r$), and total aboveground dry matter ($P_2O_5acc_t$) in spring lettuce. Confidence level: 95%.

Dependent Variable	Lsmeans ($\pm$SE)		
	INT	ORG	ORG+
$Nconc_y$ (g 100 g^{-1}) 2015	3.24 (0.10) a	2.74 (0.10) b	3.23 (0.10) a
$Nconc_y$ (g 100 g^{-1}) 2016	3.06 (0.10) a	2.23 (0.10) b	2.07 (0.10) b
$Nconc_y$ (g 100 g^{-1}) 2017	3.06 (0.10) a	2.99 (0.10) a	2.65 (0.10) b
$Nconc_r$ (g 100 g^{-1}) 2015	2.35 (0.08) a	1.92 (0.08) b	2.29 (0.08) a
$Nconc_r$ (g 100 g^{-1}) 2016	2.29 (0.08) a	1.71 (0.08) b	1.46 (0.08) c
$Nconc_r$ (g 100 g^{-1}) 2017	2.05 (0.08) a	1.99 (0.08) a	1.88 (0.08) a
$Nacc_y$ (kg ha^{-1}) 2015	35.24 (3.20) a	29.79 (3.20) a	17.69 (3.20) b
$Nacc_y$ (kg ha^{-1})2016	49.75 (3.20) a	34.12 (3.20) a	7.83 (3.20) b
$Nacc_y$ (kg ha^{-1})2017	36.01 (3.20) a	33.12 (3.20) a	14.79 (3.20) b
$Nacc_r$ (kg ha^{-1})2015	0.59 (0.04) a	0.43 (0.04) b	0.24 (0.04) c
$Nacc_r$ (kg ha^{-1})2016	1.25 (0.04) a	0.78 (0.04) b	0.17 (0.04) c
$Nacc_r$ (kg ha^{-1})2017	0.69 (0.04) a	0.67 (0.04) a	0.23 (0.04) b

Table 7. *Cont.*

Dependent Variable	Lsmeans (±SE)		
	INT	**ORG**	**ORG+**
$Nacc_t$ (kg ha^{-1})2015	49.17 (3.68) a	38.16 (3.68) b	23.15 (3.68) c
$Nacc_t$ (kg ha^{-1})2016	78.49 (3.68) a	47.53 (3.68) b	10.28 (3.68) c
$Nacc_t$ (kg ha^{-1})2017	50.13 (3.68) a	46.53 (3.68) a	19.04 (3.68) b
$Pconc_y$ (g 100 g^{-1}) 2015	0.39 (0.02) a	0.48 (0.02) b	0.45 (0.02) b
$Pconc_y$ (g 100 g^{-1}) 2016	0.49 (0.02) a	0.40 (0.02) b	0.42 (0.02) b
$Pconc_y$ (g 100 g^{-1}) 2017	0.44 (0.02) a	0.45 (0.02) a	0.44 (0.02) a
$Pconc_r$ (g 100 g^{-1}) 2015	0.28 (0.02) a	0.25 (0.02) a	0.32 (0.02) b
$Pconc_r$ (g 100 g^{-1}) 2016	0.24 (0.02) a	0.23 (0.02) a	0.29 (0.02) b
$Pconc_r$ (g 100 g^{-1}) 2017	0.26 (0.02) a	0.24 (0.02) a	0.31 (0.02) b
$P_2O_5acc_y$ (kg ha^{-1})2015	9.74 (1.20) a	11.96 (1.20) b	5.66 (1.20) c
$P_2O_5acc_y$ (kg ha^{-1})2016	18.09 (1.20) a	14.11 (1.20) b	3.69 (1.20) c
$P_2O_5acc_y$ (kg ha^{-1})2017	11.95 (1.20) a	11.28 (1.20) a	5.52 (1.20) b
$P_2O_5acc_r$ (kg ha^{-1})2015	3.84 (0.36) a	2.45 (0.36) b	1.73 (0.36) b
$P_2O_5acc_r$ (kg ha^{-1})2016	6.97 (0.36) a	4.20 (0.36) b	1.18 (0.36) c
$P_2O_5acc_r$ (kg ha^{-1})2017	4.13 (0.36) a	3.76 (0.36) a	1.59 (0.36) b
$P_2O_5acc_t$ (kg ha^{-1})2015	13.58 (1.20) a	14.41 (1.20) a	7.39 (1.20) b
$P_2O_5acc_t$ (kg ha^{-1})2016	25.07 (1.20) a	18.30 (1.20) b	4.87 (1.20) c
$P_2O_5acc_t$ (kg ha^{-1})2017	16.08 (1.20) a	15.04 (1.20) a	7.11 (1.20) b

Means followed by different letters are statistically different (95% confidence interval).

3.3.4. Summer Lettuce

For the lettuce crop grown in the summer before savoy cabbage, the statistical analysis gave significant results for all the parameters, except HI (as affected by the cropping system), $Pconc_y$, $Pconc_r$, and $P_2O_5acc_t$ (Table S4).

As for the marketable yield (expressed as fresh matter or dry matter), in 2014, INT and ORG were superior to ORG+ whereas, in 2016, there were no significant differences between ORG and ORG+ but only with INT and, in 2017, we did not find any difference among the treatments (Table 8). A similar trend was also identified for dry matter production of crop residues with the exception of 2016, when INT was higher than ORG+ only. For the total biomass of the crop, we found the same trend as for the dry matter of heads. The mean fresh weight of each lettuce head was found to be higher in INT and ORG than ORG+ in 2015, and higher in INT than ORG and ORG+ in 2016. No differences were found among treatments in 2017. The mean diameter of lettuce heads was higher in INT than ORG and higher in ORG than ORG+ in 2015 and 2016, whereas in 2017, ORG was equivalent to ORG+.

The concentration of N in the marketable yield was higher in INT than ORG and ORG+ in 2016 and 2017 (Table 9). In 2015, ORG was also higher than ORG+. The residues were richer in N in the INT plots, as well. INT was not different from ORG+ in 2015 and from ORG in 2017. The N accumulation in marketable yield was higher in INT than ORG+ in all three years. In 2015 and 2017, INT did not differ from ORG, only from ORG+. For N accumulated in crop residues, in 2015, INT and ORG were significantly higher than ORG+ whereas, in 2016, ORG+ was equivalent to ORG and, in 2017, there were no differences among the cropping systems. The total N accumulation in aboveground biomass of the lettuce was higher in INT and was significantly lower in ORG+. Nevertheless, in 2016 and 2017, the lettuce crop achieved N accumulation levels equivalent to ORG. For P concentration, summer lettuce showed normally higher levels for ORG and ORG+ with respect to INT. In 2016 and 2017, we did not find any significant differences among treatments. The residues showed a similar trend. Due to the lower biomass production, the N accumulation was very low in the ORG+ system anyway. Only in 2017 (marketable product and total biomass) and 2016 (residues), no differences were found among the three cropping systems tested.

Table 8. Least squares means and standard errors of marketable fresh yield (Y), dry matter of marketable yield (dw_y), dry matter of residues (dw_r), total aboveground dry matter (dw_t), mean fresh weight of heads (MFW), Harvest Index (HI), and mean diameter of heads (MD) in summer lettuce. Confidence level: 95%.

Dependent Variable	Lsmeans ($\pm$SE)		
	INT	ORG	ORG+
Y 2015 (Mg ha^{-1})	26.60 (2.64) a	24.54 (2.64) a	4.41 (2.64) * b
Y 2016 (Mg ha^{-1})	24.44 (2.64) a	13.79 (2.64) b	8.34 (2.64) b
Y 2017 (Mg ha^{-1})	13.08 (2.64) a	10.34 (2.64) a	7.72 (2.64) a
dw_y 2015 (Mg ha^{-1})	1.20 (0.10) a	1.12 (0.10) a	0.30 (0.10) a
dw_y 2016 (Mg ha^{-1})	1.26 (0.10) a	0.83 (0.10) b	0.54 (0.10) b
dw_y 2017 (Mg ha^{-1})	0.70 (0.10) a	0.51 (0.10) a	0.43 (0.10) a
dw_r 2015 (Mg ha^{-1})	0.87 (0.05) a	0.56 (0.05) b	0.12 (0.05) c
dw_r 2016 (Mg ha^{-1})	0.35 (0.05) a	0.25 (0.05) ab	0.18 (0.05) b
dw_r 2017 (Mg ha^{-1})	0.48 (0.05) a	0.50 (0.05) a	0.38 (0.05) a
dw_t 2015 (Mg ha^{-1})	2.07 (0.14) a	1.68 (0.14) a	0.42 (0.14) b
dw_t 2016 (Mg ha^{-1})	1.61 (0.14) a	1.08 (0.14) b	0.72 (0.14) b
dw_t 2017 (Mg ha^{-1})	1.19 (0.14) a	1.01 (0.14) a	0.81 (0.14) a
MFW 2015 (g)	266.00 (26.40) a	245.43 (26.40) a	44.14 (26.40) * b
MFW 2016 (g)	244.36 (26.40) a	137.88 (26.40) b	83.38 (26.40) b
MFW 2017 (g)	130.83 (26.40) a	103.39 (26.40) a	77.22 (26.40) a
HI 2015	0.58 (0.02) a	0.66 (0.02) b	0.71 (0.02) c
HI 2016	0.78 (0.02) a	0.77 (0.02) a	0.75 (0.02) a
HI 2017	0.59 (0.02) a	0.50 (0.02) b	0.54 (0.02) b
MD 2015 (cm)	13.13 (0.77) a	9.77 (0.77) b	5.25 (0.77) c
MD 2016 (cm)	21.61 (0.77) a	18.36 (0.77) b	13.66 (0.77) c
MD 2017 (cm)	8.64 (0.77) a	6.82 (0.77) ab	5.64 (0.77) b

Means followed by different letters are statistically different (95% confidence interval). * Value statistically not different from zero.

Table 9. Least squares means and standard errors of N concentration in marketable yield ($Nconc_y$) and residues ($Nconc_r$); N accumulation in marketable yield ($Nacc_y$), residues ($Nacc_r$), and total aboveground dry matter ($Nacc_t$); P concentration in marketable yield ($Pconc_y$) and residues ($Pconc_r$); P_2O_5 accumulation in marketable yield ($P_2O_5acc_y$), residues ($P_2O_5acc_r$), and total aboveground dry matter ($P_2O_5acc_t$) in summer lettuce. Confidence level: 95%.

Dependent Variable	Lsmeans ($\pm$SE)		
	INT	ORG	ORG+
$Nconc_y$ (g 100 g^{-1}) 2015	3.30 (0.09) a	2.92 (0.09) b	2.50 (0.09) c
$Nconc_y$ (g 100 g^{-1}) 2016	3.01 (0.09) a	2.49 (0.09) b	2.47 (0.09) b
$Nconc_y$ (g 100 g^{-1}) 2017	2.61 (0.09) a	2.34 (0.09) b	2.15 (0.09) b
$Nconc_r$ (g 100 g^{-1}) 2015	1.66 (0.08) a	2.21 (0.08) b	1.56 (0.08) a
$Nconc_r$ (g 100 g^{-1}) 2016	1.85 (0.08) a	1.37 (0.08) b	1.52 (0.08) b
$Nconc_r$ (g 100 g^{-1}) 2017	1.96 (0.08) a	1.81 (0.08) a	1.52 (0.08) b
$Nacc_y$ (kg ha^{-1}) 2015	39.47 (2.60) a	32.52 (2.60) a	7.41 (2.60) b
$Nacc_y$ (kg ha^{-1}) 2016	37.35 (2.60) a	20.57 (2.60) b	13.63 (2.60) b
$Nacc_y$ (kg ha^{-1}) 2017	18.20 (2.60) a	12.06 (2.60) ab	9.26 (2.60) b
$Nacc_r$ (kg ha^{-1}) 2015	14.55 (1.12) a	12.54 (1.12) a	1.91 (1.12) * b
$Nacc_r$ (kg ha^{-1}) 2016	6.45 (1.12) a	3.36 (1.12) ab	2.80 (1.12) b
$Nacc_r$ (kg ha^{-1}) 2017	9.40 (1.12) a	9.17 (1.12) a	5.73 (1.12) a

Table 9. *Cont.*

Dependent Variable	Lsmeans (±SE)		
	INT	**ORG**	**ORG+**
$Nacc_t$ (kg ha^{-1})2015	54.03 (3.54) a	45.06 (3.54) a	9.32 (3.54) b
$Nacc_t$ (kg ha^{-1})2016	43.81 (3.54) a	23.93 (3.54) b	16.43 (3.54) b
$Nacc_t$ (kg ha^{-1})2017	27.59 (3.54) a	21.22 (3.54) ab	14.99 (3.54) b
$Pconc_y$ (g 100 g^{-1}) 2015	0.24 (0.02) a	0.44 (0.02) b	0.47 (0.02) b
$Pconc_y$ (g 100 g^{-1}) 2016	0.37 (0.02) a	0.35 (0.02) a	0.39 (0.02) a
$Pconc_y$ (g 100 g^{-1}) 2017	0.37 (0.02) a	0.38 (0.02) a	0.37 (0.02) a
$Pconc_r$ (g 100 g^{-1}) 2015	0.15 (0.02) a	0.25 (0.02) b	0.38 (0.02) c
$Pconc_r$ (g 100 g^{-1}) 2016	0.22 (0.02) a	0.20 (0.02) b	0.26 (0.02) a
$Pconc_r$ (g 100 g^{-1}) 2017	0.24 (0.02) a	0.25 (0.02) a	0.24 (0.02) a
$P_2O_5acc_y$ (kg ha^{-1})2015	6.60 (1.03) a	11.26 (1.03) b	3.20 (1.03) c
$P_2O_5acc_y$ (kg ha^{-1})2016	10.71 (1.03) a	6.65 (1.03) b	4.89 (1.03) b
$P_2O_5acc_y$ (kg ha^{-1})2017	6.11 (1.03) a	4.61 (1.03) a	3.68 (1.03) a
$P_2O_5acc_r$ (kg ha^{-1})2015	3.11 (0.32) a	3.25 (0.32) a	1.07 (0.32) b
$P_2O_5acc_r$ (kg ha^{-1})2016	1.79 (0.32) a	1.12 (0.32) a	1.07 (0.32) a
$P_2O_5acc_r$ (kg ha^{-1})2017	2.71 (0.32) a	2.82 (0.32) a	2.03 (0.32) a
$P_2O_5acc_t$ (kg ha^{-1})2015	9.70 (1.30) a	14.51 (1.30) b	4.27 (1.30) c
$P_2O_5acc_t$ (kg ha^{-1})2016	12.49 (1.30) a	7.77 (1.30) b	5.96 (1.30) b
$P_2O_5acc_t$ (kg ha^{-1})2017	8.82 (1.30) a	7.43 (1.30) a	5.70 (1.30) a

Means followed by different letters are statistically different (95% confidence interval). * Value statistically not different from zero.

3.4. Weed Biomass at Harvest Time of the Field Vegetables

The effect of the cropping system on the dry matter produced by weeds at harvest time of savoy cabbage, fennel, and spring and summer lettuce is reported in Tables S1–S4, respectively. Only in the case of savoy cabbage, there were no significant differences among treatments. Neither were there differences due to the year. The interaction between the cropping system and year was not significant only in the case of summer lettuce.

In savoy cabbage, the organic conservative system (ORG+) did not perform worse than the other two systems in terms of weed suppression (Table 10). Only in 2016, we highlighted significantly higher weed biomass at harvest than in ORG and INT, although far under the 1 Mg ha^{-1} of dry matter. The level of weed biomass was higher in fennel in INT and ORG+ plots, whilst on average, the ORG plots showed lower values than in savoy cabbage. In one year (2014), weed biomass in ORG reached a level statistically not different from 0, resulting in a weed biomass significantly lower than INT and far lower than ORG+. In 2015, weeds were significantly more abundant in INT plots whilst ORG and ORG+ were statistically not different from each other. In 2016, we did not find any difference among the treatments, but the level of weed biomass was ca. 50% less in ORG than INT and ORG+. In the lettuce crops (i.e., spring and summer lettuce), ORG+ showed everytime higher levels of weed biomass, with only two years (i.e., 2015 for spring lettuce and 2016 for summer lettuce) with values below 1 Mg ha^{-1}. As expected, the INT system reached very low levels of weed infestation, accounting for 4 out of 6 cases for a level statistically not different from 0. The performance of ORG dramatically varied upon the lettuce crops, with significantly higher values than INT and equal to ORG+ registered in summer lettuce in 2015 and 2017. In spring lettuce, only in 2017, the ORG plots showed a mean value higher than 2 Mg ha^{-1} that was significantly higher than INT.

Table 10. Least squares means and standard errors of the dry matter of the weeds collected at harvest time of savoy cabbage, fennel, spring lettuce, and summer lettuce in the three years. Confidence level: 95%.

Dependent Variable	Lsmeans (±SE)		
	INT	**ORG**	**ORG+**
Savoy cabbage (Mg ha^{-1}) 2014	0.17 (0.14) a	1.01 (0.14) b	0.58 (0.14) c
Savoy cabbage (Mg ha^{-1}) 2015	0.36 (0.14) a	0.17 (0.14) a	0.53 (0.14) a
Savoy cabbage (Mg ha^{-1}) 2016	0.52 (0.14) a	0.29 (0.14) a	0.73 (0.14) b
Fennel (Mg ha^{-1}) 2014	0.63 (0.16) a	0.15 (0.16) * b	1.44 (0.16) c
Fennel (Mg ha^{-1}) 2015	1.41 (0.16) a	0.79 (0.16) b	0.86 (0.16) b
Fennel (Mg ha^{-1}) 2016	0.78 (0.16) a	0.39 (0.16) a	0.77 (0.16) a
Spring lettuce (Mg ha^{-1}) 2015	0.17 (0.24) * a	0.39 (0.24) * a	0.60 (0.24) b
Spring lettuce (Mg ha^{-1}) 2016	0.21 (0.24) * a	0.25 (0.24) * a	1.25 (0.24) b
Spring lettuce (Mg ha^{-1}) 2017	0.60 (0.24) a	2.08 (0.24) b	1.59 (0.24) b
Summer lettuce (Mg ha^{-1}) 2015	0.28 (0.23) * a	1.30 (0.23) b	1.28 (0.23) b
Summer lettuce (Mg ha^{-1}) 2016	0.30 (0.23) * a	0.81 (0.23) a	0.80 (0.23) a
Summer lettuce (Mg ha^{-1}) 2017	0.96 (0.23) a	1.99 (0.23) b	2.04 (0.23) b

Means followed by different letters are statistically different (95% confidence interval). * Value statistically not different from zero

3.5. Total Biomass Production and Nutrient Uptake at Crop Sequence Level

The results of the statistical analysis of the performances of the cropping systems at the level of the entire crop sequence are reported in Tables S5 and S6. As shown in Table S5, the cropping system significantly affected all the tested variables other than the yield-related ones, except the dry biomass of the weeds. The inclusion of cover crops in the analysis of the performances of the cropping system at the crop sequence level significantly affected all the tested variables, whereas the position in the crop sequence (i.e., the field) was shown to be significant only for the dry matter produced by the weeds. For yield-related variables (Table S6), the cropping systems significantly affected all the parameters whilst the position in the sequence (field) affected only the N accumulation in marketable yield.

For the total fresh marketable yield of all the crops grown in the entire crop rotation in the three years (Figure S1), overall, the INT system outperformed ORG by 12.5% and ORG+ by 161% whereas ORG was superior to ORG+ by 132%.

In Figure 2, the interaction effects between cover crops and cropping system on total production of dry matter in marketable yield (dw$_y$), residues (dw$_r$), total aboveground biomass (dw$_t$), and weeds (dw$_w$) is shown.

The total dry matter marketable yield production did not differ between INT and ORG, whereas it was lower in ORG+, whatever the level of cover crops. For total dry matter residue production, the highest value was shown by ORG CC+. ORG+ CC+ was not different from INT CC+, INT CC−, and ORG CC− but was higher than ORG+ CC−. As a result, total aboveground dry matter production of the crop sequence followed the same trend as dw$_r$. The net gains in total crop dry matter production due to inclusion of cover crops in calculations were 8.05 and 5.17 Mg ha^{-1}, respectively, for ORG and ORG+. Total weed dry matter was significantly lower in INT CC−, where the dry matter of weeds collected in the inter-crop period was not considered, than all the other treatments. The highest weed abundances were observed in INT CC+ and ORG+ CC+. Intermediate results were achieved by the remaining treatments.

In Figure 3, we reported the interaction effects between cover crops and cropping system on total N accumulation (kg N ha^{-1}) in marketable yield (Nacc$_y$), residues (Nacc$_r$), and total aboveground biomass (Nacc$_t$).

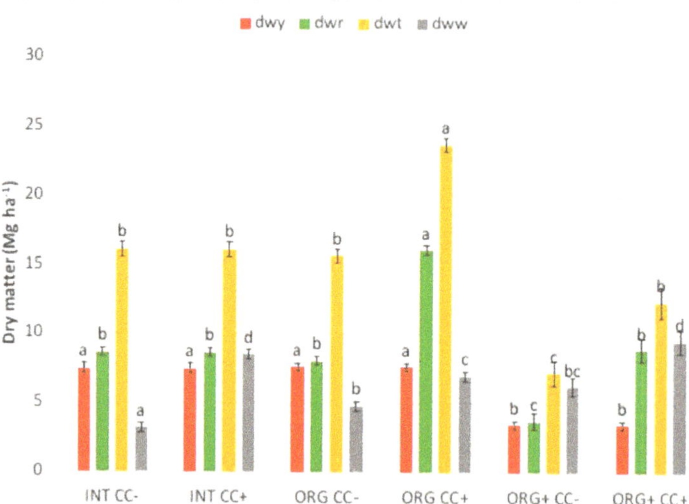

Figure 2. Interaction between cover crops (without (CC−) vs. with (CC+)) and cropping system (INT vs. ORG vs. ORG+) on dry matter production (Mg ha^{-1}) of marketable yield (dw$_y$), residues (dw$_r$), total aboveground biomass (dw$_t$), and weeds (dw$_w$) at the level of entire crop sequence: Within the same dependent variable, bars with different letters are significantly different (confidence level 0.95).

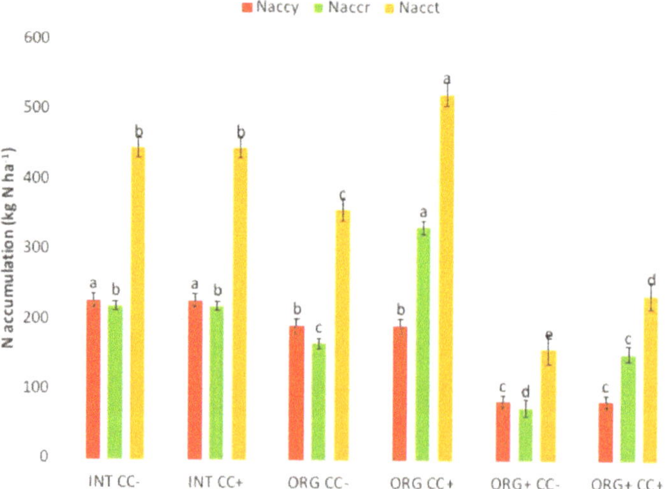

Figure 3. Interaction between cover crops (without (CC−) vs. with (CC+)) and cropping system (INT vs. ORG vs. ORG+) on total N accumulation (kg N ha^{-1}) in marketable yield (Nacc$_y$), residues (Nacc$_r$), and total aboveground biomass (Nacc$_t$) at the level of entire crop sequence: Within the same dependent variable, bars with different letters are significantly different (confidence level 0.95).

Total N accumulation in marketable product was significantly higher in INT than ORG, irrespective of cover crops level. Averaged over cover crops, ORG+ accumulated less than 100 kg N ha^{-1}, resultingly significantly lower than INT and ORG. For crop residues, we observed a different trend, with the highest N accumulation observed for ORG CC+, followed by INT CC+ and INT CC− and different from ORG CC− and ORG+ CC+. ORG CC− showed the lowest value. Total N accumulation in aboveground crop biomass followed the same trend as $Nacc_r$. ORG CC+ was the only treatment that accumulated more than 500 kg N ha^{-1}. Averaging cover crops levels, INT accounted for around 450 kg N ha^{-1} whereas the best performing ORG+ treatment (i.e., ORG+ CC+) accounted only for 236 kg N ha^{-1}. The net gain in N accumulation due to inclusion of cover crops in calculations accounted for 166 kg N ha^{-1} for ORG and 78 kg N ha^{-1} for ORG+.

3.6. Nitrogen Use Efficiency

In Table 11, the results of the analysis of N use efficiency of the single crops averaged over the three experimental years and of the entire crop sequence, considering or not considering the contribution of cover crops, are shown.

Table 11. N use efficiency indicators averaged over the three years for savoy cabbage, fennel, spring lettuce, summer lettuce, and the entire crop sequence with and without the contribution of cover crops.

Crop	Cropping System	N Budget (kg N ha^{-1})	Nsurplus$_{ti}$ * (kg N ha^{-1})	Nsurplus$_{yi}$ * (kg N ha^{-1})	NUtE$_i$ * (Mg f.m. kg^{-1} N)	NREac *	NREac$_f$ *	PFP$_i$ * (Mg f.m. kg^{-1} N)	PFP$_{fi}$ * (Mg f.m. kg^{-1} N)
Savoy cabbage	INT	−3.45	−25.78	51.90	0.18	1.03	1.24	0.18	0.22
	ORG	2.27	−57.82	0.20	0.25	1.07	1.98	0.26	0.48
	ORG+	21.79	−17.56	7.44	0.20	0.67	1.63	0.14	0.34
Fennel	INT	93.23	65.92	104.27	0.30	0.37	0.46	0.11	0.14
	ORG	81.34	34.34	65.84	0.30	0.35	0.55	0.10	0.16
	ORG+	52.60	5.23	20.99	0.40	0.32	0.81	0.13	0.34
Spring Lettuce	INT	−15.38	−38.61	−19.65	0.35	1.31	2.43	0.47	0.88
	ORG	−4.78	−28.63	−16.81	0.42	1.11	2.46	0.46	1.02
	ORG+	5.61	−20.09	−15.97	0.40	0.78	-	0.28	-
Summer Lettuce	INT	14.53	4.22	14.33	0.50	0.75	0.91	0.38	0.46
	ORG	−19.13	−29.91	−21.57	0.53	2.79	-	1.52	-
	ORG+	−1.74	−13.11	−9.61	0.51	1.19	-	0.62	-
Crop sequence without cover crops	INT	539.98	226.96	226.96	0.29	0.58	0.99	0.17	0.28
	ORG	355.58	42.55	42.55	0.32	0.65	1.53	0.21	0.49
	ORG+	313.27	0.25	0.25	0.32	0.40	1.89	0.12	0.59
Crop sequence with cover crops	INT	539.98	226.96	226.96	0.29	0.58	0.99	0.17	0.28
	ORG	456.46	42.55	42.55	0.22	0.81	2.24	0.18	0.49
	ORG+	365.30	0.25	0.25	0.21	0.53	2.81	0.11	0.59

* N surplus calculated on N accumulation in total crop biomass (Nsurplus$_{ti}$) and N accumulation in marketable yield (Nsurplus$_{yi}$); N utilization efficiency (NUtE$_i$); N Recovery Efficiency of total N inputs (NREac$_i$) and of fertilizers only (NREac$_{fi}$); and Partial Factor Productivity of total N inputs (PFP$_i$) and of fertilizers only (PFP$_{fi}$).

For savoy cabbage, the N budget (i.e., the difference between all the N inputs and N accumulation in total biomass) was positive only for ORG and ORG+. In particular, ORG+ resulted in the lowest value, with about 22 kg N ha^{-1} of surplus. Overall, for cabbage, the three systems did not overconsume or exploit N. Nevertheless, the important contribution of N from sources other than fertilizers was clearly shown by the negative values of N surplus calculated in terms of total N accumulation for all three systems (Nsurplus$_{ti}$). The fertilizers covered actually the N accumulation of corymbs in ORG and ORG+ whilst gave a surplus of around 50 kg N ha^{-1} in INT (Nsurplus$_{yi}$). Apparently, the efficiency in converting into marketable yield the unit of N accumulated in the biomass was not different among the systems (NUtE) and accounted for around 0.2 Mg f.m. kg^{-1} total N accumulation. The recovery of total N inputs was close to 1 (i.e., the level at which N accumulated in total biomass was equal to the N inputs) for INT and ORG, whilst ORG+ accumulated only 67% of total N inputs. If considering only N from fertilizers, the three systems clearly all showed they accumulated also N from other sources, as they all showed values far higher than 1. The efficiency in converting the unit of N supplied in marketable yield (PFP) was higher in the ORG system, either considering the totality of N inputs or only the fertilizers. Interestingly, ORG+ outperformed INT when considering only N from fertilizers as an input.

For fennel, the N budget was sensibly more positive than for savoy cabbage. The ORG+ revealed an N surplus close to zero when considering the total N accumulation (only 5.23 kg N ha^{-1}). The NUtE was slightly higher in the ORG+ and converted more efficiently the N accumulated into swollen bases (+0.1 Mg f.m. kg^{-1} N). The PFP was lower than in savoy cabbage and reached the maximum in ORG+.

For spring lettuce, ORG+ was the only treatment showing a slightly positive N budget (5.61 kg N ha^{-1}), but when considering as N inputs, only the N from fertilization of all the treatments gave negative values, meaning N outputs were higher than inputs due to low values of N from fertilizers. The NUtE results did not show any difference among the systems and averaged around 0.4 Mg f.m. kg^{-1} N. The lettuce in ORG+ plots did not uptake 22% of the N supplied as total inputs. NReac$_{\mathrm{fi}}$ and PFP$_{\mathrm{fi}}$ were not calculated for ORG+ as N fertilizers were not applied. More than double the N accumulated in crop biomass in ORG and INT came from sources other than fertilizers (NReac$_{\mathrm{fi}}$). ORG+ was the less efficient system in terms of conversion of N supplied into marketable yield.

In summer lettuce, the INT system resulted in a positive N budget (+14.53 kg N ha^{-1}) and surplus (4.22 and 14.33 kg N ha^{-1}, respectively, for Nsurplus$_{\mathrm{ti}}$ and Nsurplus$_{\mathrm{fi}}$) whereas ORG and ORG+ always gave negative values due to nonuse of fertilizers. The NUTe was not different among the systems and reached the highest values in the crop rotation (around 0.5 Mg f.m. kg^{-1} N). ORG (2.79) and ORG+ (1.19) showed the highest efficiency in recovery of N supplied as total inputs, whilst INT did not reach the tie value of 1 even when considering only N from fertilization. The productivity of N units (PFP) was higher in ORG than ORG+ and then INT.

Considering the entire crop sequence, it is clear how all the systems produced high N surplus expressed as N budget that peaked 540 kg N ha^{-1} in INT, 356 kg N ha^{-1} in ORG, and 313 kg N ha^{-1} in ORG+. If considering also N fixation of legume cover crops, the N budget of ORG and ORG+ reached, respectively, 457 and 365 kg N ha^{-1}. Interestingly, the two organic systems differed from INT in terms of N surplus that was close to 0 but still positive for INT and very negative for ORG and ORG+, especially when considering also N accumulated by cover crops, as we did not distinguish between N accumulation derived from N fixation. This means the two organic systems strongly relied on N sources other than fertilizers. When considering only N accumulation in marketable yield, the N surplus was close to 0 for ORG+, positive for ORG (around 43 kg N ha^{-1}), and still high for INT (227 kg N ha^{-1}). Averaged over crops, NUtE was around 0.30 Mg f.m. kg^{-1} N for all the systems when not considering cover crops, whereas it became 0.1 Mg f.m. kg^{-1} N lower in ORG and ORG+ when including N from cover crops in calculations. The N recovery was far lower from 1 in all the systems when considering total N inputs, with ORG showing the highest value (0.65 Mg f.m. kg^{-1} N). Including N accumulated by cover crops increased the efficiency of ORG and ORG+, with ORG reaching 0.81 Mg f.m. kg^{-1} N. If accounting only N from fertilization, the results clearly showed how INT was able to accumulate 99% of fertilizer N whilst ORG and ORG+ were underfertilized and relied upon additional N from other natural sources. Finally, the PFP of total N inputs was comparable among the systems and a bit lower in ORG+ than ORG and INT. Nevertheless, the PFP of fertilizers only clearly segregated among INT and the two organic systems. Due to the low N fertilization rates, ORG+ resulted in being the most productive system per unit of N supplied as fertilizers (+0.1 Mg f.m. kg^{-1} N with respect to ORG and +0.31 Mg f.m. kg^{-1} N with respect to INT).

4. Discussion

In this work, we studied the agronomic performances of an organic conservative management of a two-year field vegetable crop rotation compared to a standard organic and an integrated management system in Mediterranean conditions.

Our study confirms previous evidences (e.g., References [18,27,28,39]) that organic no-till systems are promising strategies to improve the sustainability of organic field vegetable systems but still need strong development and further investigations. In our experiment, we designed the organic conservative system with the main aims to reduce GHG emissions, to produce nonrenewable energy saving, and to emphasize use of internal natural resources in compliance with the principles of organic no-till [6,20] and agroecology [27,39]. This resulted in very basic application of organic fertilizers in the ORG+ plots, where most of the regulating services (i.e., nutrient availability and weed suppression) were supposed to be provided by the cover crops grown as living or dead mulch. Given the not so high and stable biomass production of the dead mulch and the variable growth of the living mulch grown in the ORG+ plots (Table 1), the provision of agroecological services was not expressed enough to enhance system performances. Further research efforts are still needed to identify the cover crop species and management options most adapted to no-till conditions in order to enhance the level and the stability of ecosystem service delivery by cover crops. Selection of cover crop species should be done taking into account traits related to rooting capacity, high nutrient uptake and mobilization, N_2 symbiotic fixation, low water consumption, quick soil cover, and creeping habitus (especially for living mulch use) [20]. The level of biomass production of the cover crops is indeed the crucial factor behind the functioning of no-till, cover crop-based cropping systems, as also pointed out by Reference [28].

For savoy cabbage, our results clearly showed how lack of nitrogen was likely the most important limiting factor for crop yield and N accumulation in ORG+. In the best years (2014 and 2015), the cabbage managed under ORG+ conditions yielded 50% lower than INT, whereas it was almost unable to complete the reproductive phase in 2016 (Table 2). Nevertheless, the results achieved by the ORG and INT systems were overall far below the standard for Central Italy [40] but in line with other similar experiments [41]. It is noteworthy that the harvest index was normally higher in ORG+ plots than in INT (Table 2). This result, combined with the low dry matter production of corymbs, highlights that also the vegetative growth was not well completed by the crop in the conservative system, likely due to a lack of readily available nitrogen from the initial stages after transplanting. To prove this hypothesis, further investigations are needed, looking at early indicators of crop nutrient status at the vegetative stage, such as the NDVI (Normalized Difference Vegetation Index) or the LAI (Leaf Area Index), as suggested by Reference [42].

Besides the concentration of total N in the root zone, the mineralization rate of soil organic matter and of organic fertilizers applied uniquely at the transplanting stage also might have differed between ORG+ and tilled systems. In the ORG+ system, the N fertilizers were broadcast spread over the soil just before transplanting the cabbage and this might have likely caused poor contact with the soil and consequently a slowdown in the mineralization rate of the fertilizers. The N use efficiency indicators studied (Table 11), in particular, the NReacc (showing a 33% of reduced recovery of N applied as fertilizers in ORG+), clearly support this hypothesis. In a recent paper published on GHG emission in the same experiment [26], a lower N_2O emission from ORG+ plots than in INT and ORG was demonstrated. This might have been due to the low supply of fertilizers in the organic conservative system. Nevertheless, in the same paper, peaks of N_2O emission after application of organic fertilizers have been reported, possibly due to no incorporation of the fertilizers. Our findings thus support the option to increase N fertilization rates at least in the transition phase to no-till in order to better support plant growth, given the uncertainty of the mineralization rate of the organic fertilizers in untilled soils. Another important option to be tested is to place fertilizers directly into the crop furrow [43] on the transplanting date in order to enhance contact between nutrients and roots and, more importantly, to stimulate the mineralization of the organic material and fertilizer N uptake while preventing high N_2O emissions. Given the organic nature of the fertilizers and their low N concentration, a significant caustic effect of N on crop roots should not occur with in-furrow applications.

Among other potential stressors, excluding any effect of noxious organisms (e.g., pests and diseases), we can argue that also soil compaction caused by no-till in first 0–10 cm soil layer might have played a role. This is well known in literature on no-till, especially in the transitional stage from inversion tillage to conservative management as in our case [6,44]. The use of the modified transplanting machine mounting shank openers [32] should have been reduced at least at the beginning problems of compaction for rootlets. At later stages, the persistence of shallow compaction might have caused a limited vertical root development that we were not able to assess.

Actually, neither weeds nor P seemed to have been the real limiting factors for cabbage. For weed biomass at harvest (Table 10), we did not find clear and strong differences between ORG+ and tilled systems. Likewise, for P concentration in plant tissues (Table 3), we did not observe significant depletion in ORG+ plots, but rather often an increase. This trend was also evident for fennel (Table 5), for which we did not detect any significant differences between ORG+ and INT. This is an interesting issue that is worth further investigating in the future. We hypothesize that, besides a concentration effect due to poorer crop biomass produced, there might have been other reasons for this increase in P content under no-till plus living mulch, first of all symbiosis with arbuscular mycorrhizal fungi (AMF) and soil acidification mediated by the living mulch root exudates [45].

Nevertheless, besides P availability, the contribution of the studied living mulch of red clover was not appreciable in terms of crop advantages. Other authors [46] highlighted how the management of living mulch plays a key role in determining its ability to grow without competing for resources with the crop and to deliver weed suppression and nutrient mobilization. In particular, it was shown how a sowing date of the living mulch earlier than the transplanting date of the vegetable crop can reduce the performance of the living mulch compared to sowing contemporary to transplanting [46]. In our case, the red clover did not always establish well after its direct seeding due to soil compaction and poor seed-soil contact. Then, the clover covered the soil pretty well at crop transplanting (summer lettuce and cabbage). Anyway, in many cases, we observed the clover outgrowing the cabbage and, more often, summer lettuce at early development stages of the crops, whereas competition with weeds at later stages was not satisfactory. This was also because the in-crop management of the living mulch, i.e., inter-row flaming, was feasible and effective only until the crop did not cover the rows, but in many cases, the most aggressive weeds started to grow only later in these stages. This was especially the case of summer lettuce, for which we observed repeatedly problems of summer grass weed species (e.g., *Digitaria sanguinalis* (L.) Scop., *Setaria italica* subsp., and *viridis* (L.) Beauv.) escaping from control with flaming. Another weakness of this management system was the control of the weeds within the row that was simply not feasible with the operating machines available. Developing machinery for effective weed and living mulch management also at later stages and within the row is thus required to improve no-till systems based on living mulch. To enhance the living mulch establishment, as its direct seeding can be problematic, and to reduce soil compaction due to passes of heavy machinery like direct drillers, alternative strategies, e.g., testing permanent living mulch lasting for years before regeneration as suggested by References [6,47], should be tested. This could also help solve the problem of controlling weeds within the crop row as creeping, permanent living mulch can have enough time to cover the entire soil surface during their growth.

For fennel and, more importantly, on the two lettuce crops, we identified in weed management the most important factor together with nutrient management affecting crop yield in the organic conservative system. As pointed out by Reference [48], weed control represents an important yield determinant in organic fennel due to scarce competitiveness of the crop. For lettuce, weed competition is a major issue given its short cycle, small plant height, low soil cover capacity, and shallow root [49].

Weed biomass at harvest in spring lettuce was always higher than the other systems except for in 2016 (not different from ORG) (Table 10). In summer lettuce, ORG+ had more weeds than INT in all years, but not significantly in 2015. ORG+ did not differ from ORG in all three years (Table 10). The high weed presence and the low yield observed in summer lettuce might be partially explained by the performances of the living mulch of red clover. The lowest yield observed in 2015 (4.41 Mg f.m. ha^{-1}, a value not statistically different from 0) can be linked, in our opinion, to the overgrowth of the living mulch, which peaked 2.27 Mg d.m. ha^{-1} (Table 1) and was too competitive with the crop from early stages. In 2016, when the summer lettuce reached the highest yield (Table 8), the biomass production of the living mulch was almost null. Nevertheless, the weeds were also not so aggressive (Table 10), maybe because weather conditions more favorable to the crop. In 2017, when the weed biomass reached a peak over 2 Mg d.m. ha^{-1} (Table 10), the crop yield was not much depleted, maybe because of favorable weather conditions and also a slightly higher N availability from the living mulch, the biomass of which was a bit higher than in 2016 (Table 1).

For spring lettuce, the relationships between crop yield and weed biomass were more evident than in summer lettuce, as also shown by the huge yield depletion in the ORG system in 2017 in the presence of the highest weed abundance (Tables 6 and 10). The absence of a new cover crop grown immediately before the spring lettuce and the poor regeneration of the red clover after cabbage harvest might have led to insufficient weed control.

In our experiment, the weed biomass at harvest of fennel was higher in ORG+ than ORG and INT only in 2014 (Table 10), leading to a marketable yield much lower than achievable [40,41] (Table 4). The yield was still low in 2016, although weed biomass was half that of 2014 (Table 10). Nevertheless, in 2015, the marketable yield of fennel in the conservative system reached a peak and did not differ from the standard organic system (ORG) (Table 4). If we look at the composition and biomass produced by the summer cover crop mixture grown before fennel (Table 1), we can easily argue that the performances of the fennel were very related to the growth of the cover crops grown before its transplant and terminated as dead mulch. In 2014, the low yield of fennel and the high weed biomass at harvest can be explained by the low biomass produced by the dead mulch (only 1.31 Mg d.m. ha^{-1}). For dead mulch, the importance of achieving high amounts of biomass production of cover crops to produce good soil cover, weed suppression, and nutrient release has been reported by many authors [6,50] and, in a recent paper, produced in similar conditions [28]. The summer cover crop mixture adopted in this study performed quite well but with the contribution of 2–3 species, whilst one of them (i.e., buckwheat) was very scarce in the canopy (Table 1). This finding emphasizes the need to investigate further in species/varieties of cover crops adapted to use as dead mulches in no-till systems. Besides quantity, also quality of cover crop biomass can be a key factor in terms of service delivery. In 2015 and 2016, when the cover crop biomass production was satisfactory, we can identify two different compositions of the mixture, with grass species (i.e., foxtail millet and grain millet) dominating in 2015 and a more balanced composition in 2016 (Table 1). This difference might have led to different killing rates and different kinetics of decomposition [31]. The termination technique (i.e., roller crimping plus flaming [28,31]) was very effective in both years in terms of killing rate, which was proximate to 95%. In 2015, the more abundant dead mulch obtained allowed the fennel to complete regularly the first stages after transplanting and to establish well, given the good soil moisture level conserved below the mulch and the thickness of the mulch, which prevented weeds from emerging at early stages. The not-so-quick mineralization rate of the biomass, characterized by a dominance of grass plants (i.e., high C:N ratio), avoided quick disappearance of the mulch that was as effective as at early stages in reducing weed emergence and growth, especially at later stages. Probably, this did not happen in 2016, when a higher presence of red cowpea biomass in the mixture might have led to a quicker decomposition of the dead mulch, freeing space for weeds to develop earlier than in 2015. In fennel, inter-row flaming was not as effective as in cabbage due to the presence of the dead mulch layer, which did not allow to enhance exposure time because of the burning risk. Thus, our findings encourage further research efforts aimed at identification and testing in different pedoclimatic conditions of high biomass producing cover crops with high long-lasting capacity. Different termination techniques and machinery should be also tested in order to allow for distribution of the dead mulch along the crop furrow. Monitoring of mineralization rates of dead mulch provided by different cover crop species managed differently is also recommended.

As shown in Table 5, in 2016, the concentration of N in swollen bases and residues of fennel in ORG+ reached the lowest values, revealing insufficient crop N uptake and a nonrelevant contribution from the cover crops in terms of N supply. Red cowpea, the only legume in the mixture, which was supposed to deliver N to fennel through N_2 biological fixation, did not produce nodules in the first year, maybe due to low presence of the required *Rhizobium* strain, whereas it showed regularly root nodules in the second and third year. This behavior might have produced different interactions with the grass companion cover crops (foxtail millet and grain millet) in terms of N availability, leading to different levels and quality of service provision to the fennel.

In terms of nutrient management, for fennel and cabbage, our findings highlighted how a unique application of N fertilizers at crop transplant could not be enough to sustain the crop during its growth. Fertigation with organic soluble fertilizers can be an option to achieve an increased nutrient availability for the vegetables, to simultaneously reduce water volumes for irrigation, and to reduce water availability for weeds by concentrating irrigation on the crop row [51]. Nevertheless, this option may interfere with mechanical/thermal weed control due to the presence of the irrigation hoses on topsoil. Subirrigation combined with no-till can be an alternative valuable option in that sense [51].

Overall, from an agroecological point of view, our results demonstrate that the total biomass production of the low-input organic conservative systems can be as high as in the standard organic and integrated systems if cover crop biomass is also considered. Cover crops thus were confirmed to be indispensable tools in conservative low-input systems. What clearly made a huge difference was the proportion of marketable yield on total biomass, which was normally higher in the tilled systems because of faster mineralization of crop residues and organic fertilizers and lower weed abundance. Insisting on the fine-tuning of organic conservation systems is thus worth to be pursued in order to enhance the marketable productivity of these systems.

5. Conclusions

The organic conservative system tested in our experiment (ORG+) revealed to be ineffective in terms of crop yield and N uptake for the four vegetable crops. Nevertheless, potentialities in terms of reduction of environmental pollution risks by avoiding nitrogen surplus in the soil and better exploitation of natural internal resources (N from N_2-fixation and higher availability of soil P) were highlighted. As organic cover crop and no-till based cropping systems should express their potential when designed tailored to local pedoclimatic and agronomic conditions, we encourage further development of the system through additional investigations on soil nutrient cycling processes and weed dynamics in no-till systems, as well as on developing and testing innovative technologies for the management of cover crops, weeds, irrigation, and fertilization adapted to such systems.

Supplementary Materials: The following are available online at http://www.mdpi.com/2073-4395/9/12/810/s1, Figure S1: Total fresh marketable yield (Mg ha^{-1}) of the entire crop sequence under the three cropping systems (INT, ORG, and ORG+). Bars are standard errors. Confidence level 95%, Table S1: Agricultural practices carried out in Field 1 for each crop in the whole experimental period, Table S2: Agricultural practices carried out in Field 2 for each crop in the whole experimental period, Table S3: Type III analysis of variance table with Satterthwaite's method of the model lmer (y~cropping system*year+(1|block)+(1|year) for savoy cabbage. Confidence level: 95%, Table S4: Type III analysis of variance table with Satterthwaite's method of the model lmer (y~cropping system*year+(1|block)+(1|year) for fennel. Confidence level: 95%, Table S5: Type III analysis of variance table with Satterthwaite's method of the model lmer (y~cropping system*year+(1|block)+(1|year) for spring lettuce. Confidence level: 95%, Table S6: Type III analysis of variance table with Satterthwaite's method of the model lmer (y~cropping system*year+(1|block)+(1|year) for summer lettuce. Confidence level: 95%, Table S7: Field log of Field 1 and Field 2 with dates and operational details on each field operation practiced in the three years of the experiment.

Author Contributions: Conceptualization, D.A., M.F., and C.F.; methodology, D.A., C.F., and L.M.; validation, D.A., C.F., and M.M.; formal analysis, L.M.; investigation, D.A., M.S., C.F., and L.A.C.; resources, C.F., M.M., A.P., and M.R.; data curation, D.A., M.S., and C.F.; writing—original draft preparation, D.A. and L.M.; writing—review and editing, D.A., C.F., M.M., A.P., L.A.C., M.F., and M.R.; project administration, C.F.; funding acquisition, C.F.; statistical analysis of data, L.M.

Funding: This research was carried out within the project SMOCA "Smart Management of Organic Conservation Agriculture" (http://smoca.agr.unipi.it/) funded by the Italian Ministry of University and Research (MIUR) within the program FIRB-2013 (Future in Research) and MIUR-FIRB13 (project number: RBFR13L8J6).

Acknowledgments: The authors would like to acknowledge the staff at the "Enrico Avanzi" Centre for Agro-Environmental Research of the University of Pisa who managed the field trials and provided technical support throughout. In particular, we are grateful to Alessandro Pannocchia, Giovanni Melai, Marco Della Croce, and Paolo Gronchi, who were in charge of field operations. We also thank Roberta Del Sarto, Nadia Ceccanti,

Rosenda Landi, and Serena Sbrana for their help in sample processing. Finally, we acknowledge also Rosalba Risaliti and Sabrina Ciampa for their support in the chemical analysis of plant samples.

Conflicts of Interest: The authors declare no conflict of interest.

References

1. Bommarco, R.; Kleijn, D.; Potts, S. Ecological intensification: Harnessing ecosystem services for food security. *Trends Ecol. Evol.* **2013**, *28*, 230–238. [CrossRef] [PubMed]
2. Bedoussac, L.; Journet, E.P.; Hauggaard-Nielsen, H.; Naudin, C.; Corre-Hellou, G.; Jensen, E.S.; Prieur, L.; Justes, E. Ecological principles underlying the increase of productivity achieved by cereal-grain legume intercrops in organic farming. A review. *Agron. Sustain. Dev.* **2015**, *35*, 911–935. [CrossRef]
3. Mäder, P.; Fliessbach, A.; Dubois, D.; Gunst, L.; Fried, P.; Niggli, U. Soil fertility and biodiversity in organic farming. *Science* **2002**, *296*, 1694–1697. [CrossRef] [PubMed]
4. Gattinger, A.; Muller, A.; Haeni, M.; Skinner, C.; Fliessbach, A.; Buchmann, N.; Mader, P.; Stolze, M.; Smith, P.; El-Hage Scialabba, N.; et al. Enhanced top soil carbon stocks under organic farming. *Proc. Natl. Acad. Sci. USA* **2012**, *109*, 18226–18231. [CrossRef] [PubMed]
5. Lernoud, J.; Willer, H. The World of Organic Agriculture. Statistics and Emerging Trends 2019. Research Institute of Organic Agriculture (FiBL), Frick, and IFOAM-Organics International, Bonn. 2019. Available online: https://shop.fibl.org/CHen/mwdownloads/download/link/id/1202/?ref=1 (accessed on 15 November 2019).
6. Peigné, J.; Ball, B.C.; Roger-Estrade, J.; David, C. Is conservation tillage suitable for organic farming? A review. *Soil Use Manage.* **2007**, *23*, 129–144. [CrossRef]
7. Morris, D.R.; Gilbert, R.A.; Reicosky, D.C.; Gesch, R.W. Oxidation potentials of soil organic matter in histosols under different tillage methods. *Soil Sci. Soc. Am. J.* **2004**, *68*, 817–826. [CrossRef]
8. Grandy, A.S.; Robertson, G.P.; Thelen, K.D. Do productivity and environmental trade-offs justify periodically cultivating no-till cropping systems? *Agron. J.* **2006**, *98*, 1377–1383. [CrossRef]
9. Mazzoncini, M.; Antichi, D.; Di Bene, C.; Risaliti, R.; Petri, M.; Bonari, E. Soil carbon and nitrogen changes after 28 years of no-tillage management under Mediterranean conditions. *Eur. J. Agron.* **2016**, *77*, 156–165. [CrossRef]
10. Johnson, J.M.F.; Franzluebbers, A.J.; Weyers, S.L.; Reicosky, D.C. Agricultural opportunities to mitigate greenhouse gas emissions. *Environl. Pollut.* **2007**, *150*, 107–124. [CrossRef]
11. Carr, P.; Gramig, G.; Liebig, M. Impacts of organic zero tillage systems on crops, weeds, and soil quality. *Sustainability* **2013**, *5*, 3172–3201. [CrossRef]
12. Gadermaier, F.; Berner, A.; Fließbach, A.; Friedel, J.; Mäder, P. Impact of reduced tillage on soil organic carbon and nutrient budgets under organic farming. *Renew. Agric. Food Syst.* **2011**, *27*, 68–80. [CrossRef]
13. Casagrande, M.; Peigné, J.; Payet, V.; Mäder, P.; Sans, F.X.; Blanco-Moreno, J.M.; Antichi, D.; Bàrberi, P.; Beeckman, A.; Bigongiali, F.; et al. Organic farmers' motivations and challenges for adopting conservation agriculture in Europe. *Organ. Agric.* **2015**, *6*, 281–295. [CrossRef]
14. Food and Agriculture Organization of the United Nations (FAO). Available online: http://www.fao.org/conservation-agriculture/en/ (accessed on 16 November 2019).
15. Blanco-Canqui, H.; Shaver, T.M.; Lindquist, J.L.; Shapiro, C.A.; Elmore, R.W.; Francis, C.A.; Hergert, G.W. Cover crops and ecosystem services: Insights from studies in temperate soils. *Agron. J.* **2015**, *107*, 2449–2474. [CrossRef]
16. Dabney, S.M.; Delgado, J.A.; Reeves, D.W. Using winter cover crops to improve soil and water quality. *Commun. Soil Sci. Plan.* **2001**, *32*, 1221–1250. [CrossRef]
17. Thorup-Kristensen, K.; Magid, J.; Jensen, L.S. Catch crops and green manures as biological tools in nitrogen management in temperate zones. *Adv. Agron.* **2003**, *79*, 227–302.
18. Diacono, M.; Ciaccia, C.; Canali, S.; Fiore, A.; Montemurro, F. Assessment of agro-ecological service crop managements combined with organic fertilisation strategies in organic melon crop. *Ital. J. Agron.* **2018**, *13*, 172–182. [CrossRef]
19. Wittwer, R.; Dorn, B.; Jossi, W.; van der Heijden, M. Cover crops support ecological intensification of arable cropping systems. *Sci. Rep.* **2017**, *7*, 41911. [CrossRef]

20. Vincent-Caboud, L.; Peigné, J.; Casagrande, M.; Silva, E. Overview of organic cover crop-based no-tillage technique in Europe: Farmers' practices and research challenges. *Agriculture* **2017**, *7*, 42. [CrossRef]

21. Nichols, V.; Verhulst, N.; Cox, R.; Govaerts, B. Weed dynamics and conservation agriculture principles: A review. *Field Crop. Res.* **2015**, *183*, 56–68. [CrossRef]

22. Van Den Bossche, A.; De Bolle, S.; De Neve, S.; Hofman, G. Effect of tillage intensity on N mineralization of different crop residues in a temperate climate. *Soil Till. Res.* **2009**, *103*, 316–324. [CrossRef]

23. Baggs, E.M.; Watson, C.A.; and Rees, R.M. The fate of nitrogen from incorporated cover crop and green manure residues. *Nutr Cycl. Agroecosyst.* **2000**, *56*, 153–163. [CrossRef]

24. Rochette, P.; Angers, D.; Chantigny, M.; Bertrand, N. Nitrous oxide emissions respond differently to no-till in a loam and a heavy clay soil. *Soil Sci. Soc. Am. J.* **2008**, *72*, 1363–1369. [CrossRef]

25. Van Kessel, C.; Venterea, R.; Six, J.; Adviento-Borbe, M.A.; Linquist, B.; van Groenigen, K.J. Climate, duration, and N placement determine N2O emissions in reduced tillage systems: A meta-analysis. *Glob. Chang. Biol.* **2013**, *19*, 33–44. [CrossRef] [PubMed]

26. Bosco, S.; Volpi, I.; Antichi, D.; Ragaglini, G.; Frasconi, C. Greenhouse gas emissions from soil cultivated with vegetables in crop rotation under Integrated, organic and organic conservation management in a mediterranean environment. *Agronomy* **2019**, *9*, 446. [CrossRef]

27. Canali, S.; Diacono, M.; Campanelli, G.; Montemurro, F. Organic no-till with roller crimpers: Agro-ecosystem services and applications in organic Mediterranean vegetable productions. *Sustain. Agric. Res.* **2015**, *4*, 70. [CrossRef]

28. Abou Chehade, L.; Antichi, D.; Martelloni, L.; Frasconi, C.; Sbrana, M.; Mazzoncini, M.; Peruzzi, A. Evaluation of the agronomic performance of organic processing tomato as affected by different cover crop residues management. *Agronomy* **2019**, *9*, 504. [CrossRef]

29. Soil Survey Staff. *Keys to Soil Taxonomy*, 12th ed.; USDA-Natural Resources Conservation Service: Washington, DC, USA, 2014.

30. Gómez, K.A.; Gómez, A.A. *Statistical Procedures for Agricultural Research*; John Wiley Sons: New York, NY, USA, 1984.

31. Frasconi, F.; Martelloni, L.; Antichi, D.; Raffaelli, M.; Fontanelli, M.; Peruzzi, A.; Benincasa, P.; Tosti, G. Combining roller crimpers and flaming for the termination of cover crops in herbicide-free no-till cropping systems. *PLoS ONE* **2019**, *14*, e0211573. [CrossRef]

32. Frasconi, C.; Martelloni, L.; Raffaelli, M.; Fontanelli, M.; Abou Chehade, L.; Peruzzi, A.; Antichi, D. A field vegetable transplanter for use in both tilled and no-till soils. *Trans. ASABE* **2019**, *62*, 593–602. [CrossRef]

33. Parris, K. Agricultural nutrient balances as agri-environmental indicators: An OECD perspective. *Environ. Pollut* **1998**, *102* (Suppl. S1), 219–225. [CrossRef]

34. Huggins, D.R.; Pan, W.L. Key Indicators for assessing nitrogen use efficiency in cereal-based agroecosystems. *J. Crop. Prod.* **2003**, *8*, 157–185. [CrossRef]

35. Liu, J.; Liu, H.; Huang, S.; Yang, X.; Wang, B.; Li, X.; Ma, Y. Nitrogen efficiency in long-term wheat–maize cropping systems under diverse field sites in China. *Field Crop. Res.* **2010**, *118*, 145–151. [CrossRef]

36. Cassman, K.G.; Gines, G.C.; Dizon, M.A.; Samson, M.I.; Alcantara, J.M. Nitrogen-use efficiency in tropical lowland rice systems: Contributions from indigenous and applied nitrogen. *Field Crop. Res.* **1996**, *47*, 1–12. [CrossRef]

37. Kuznetsova, A.; Brockhoff, P.B.; Christensen, R.H.B. Lmer Test: Tests in Linear Mixed Effects Models. R Package Version 2.0–32. Available online: https://CRAN.R-project.org/package=lmerTest (accessed on 25 July 2018).

38. R Core Team. R: A Language and Environment for Statistical Computing. R Foundation for Statistical Computing, Vienna, Austria. 2016. Available online: https://www.R-project.org/ (accessed on 15 November 2019).

39. Diacono, M.; Persiani, A.; Fiore, A.; Montemurro, F.; Canali, S. Agro-Ecology for potential adaptation of horticultural systems to climate change: Agronomic and energetic performance evaluation. *Agronomy* **2017**, *7*, 35. [CrossRef]

40. Tesi, R. *Orticoltura Mediterranea Sostenibile*; Patron Editore: Bologna, Italy, 2010.

41. Campanelli, G.; Canali, S. Crop Production and Environmental Effects in Conventional and Organic Vegetable Farming Systems: The Case of a Long-Term Experiment in Mediterranean Conditions (Central Italy). *J. Sustain. Agric.* **2012**, *36*, 599–619. [CrossRef]

42. Ji, R.; Min, J.; Wang, Y.; Cheng, H.; Zhang, H.; Shi, W. In-Season yield prediction of cabbage with a hand-held active canopy sensor. *Sensors* **2017**, *17*, 2287. [CrossRef]

43. Stevens, W.B.; Blaylock, A.D.; Krall, J.M.; Hopkins, B.G.; Ellsworth, J.W. Sugar beet yield and nitrogen use efficiency with preplant broadcast, banded, or point-injected nitrogen application mention of trade names, proprietary products, or specific equipment is intended for reader information only and constitutes neither a guarantee nor warranty by the ARS-USDA, nor does it imply approval of the product named to the exclusion of other products. *Agron. J.* **2006**, *99*, 1252–1259.

44. Blanco-Canqui, H.; Ruis, S.J. No-tillage and soil physical environment. *Geoderma* **2018**, *326*, 164–200. [CrossRef]

45. Deguchi, S.; Uozumi, S.; Tawaraya, K.; Kawamoto, H.; Tanaka, O. Living mulch with white clover improves phosphorus nutrition of maize of early growth stage. *Soil Sci. Plant. Nutr.* **2005**, *51*, 573–576. [CrossRef]

46. Diacono, M.; Fiore, A.; Farina, R.; Canali, R.; Di Bene, C.; Testani, E.; Montemurro, F. Combined agro-ecological strategies for adaptation of organic horticultural systems to climate change in Mediterranean environment. *Ital. J. Agron.* **2016**, *11*, 85–91. [CrossRef]

47. Pieper, J.R.; Brown, R.; Amador, J.A. Effects of three conservation tillage strategies on yields and soil health in a mixed vegetable production system. *HortScience* **2015**, *50*, 1770–1776. [CrossRef]

48. Fontanelli, M.; Frasconi, C.; Martelloni, L.; Pirchio, M.; Raffaelli, M.; Peruzzi, A. Innovative strategies and machines for physical weed control in organic and integrate vegetable crops. *Chem. Eng. Trans.* **2015**, *44*, 211–216.

49. Lati, R.; Mou, B.; Rachuy, J.; Smith, R.; Dara, S.; Daugovish, O.; Fennimore, S. Weed Management in Transplanted Lettuce with Pendimethalin and S-Metolachlor. *Weed Technol.* **2015**, *29*, 827–834. [CrossRef]

50. Morse, R.D. No-till vegetable production—Its time is now. *HortTechnol* **1999**, *9*, 373–379. [CrossRef]

51. Lai, R. Managing Soils for Food Security and Climate Change. *J. Crop Improv.* **2007**, *19*, 49–71. [CrossRef]

Article

A Technical-Economic Comparison between Conventional Tillage and Conservative Techniques in Paddy-Rice Production Practice in Northern Italy

Aldo Calcante * and Roberto Oberti

Department of Agricultural and Environmental Sciences, Università degli Studi di Milano, Via Celoria 2,
IT-20133 Milan, Italy; roberto.oberti@unimi.it
* Correspondence: aldo.calcante@unimi.it; Tel.: +39-02-503-16465

Received: 22 November 2019; Accepted: 11 December 2019; Published: 13 December 2019

Abstract: In this study a technical-economic comparison was conducted to compare three different agronomic practices applied to paddy rice cultivation areas in Italy: one based on conventional tillage (CT), and two adopting conservative agriculture approaches, namely minimum tillage (MT) and no-tillage (NT). Data about production inputs (seed, fertilizers, agrochemicals, fuel) and working time were measured for each technique during the whole production season in three experimental fields. The total production costs were computed by adding the mechanization costs, calculated through the ASABE (American Society of Agricultural and Biological Engineers) EP (Engineering Practice) 496.3 methodology, and the production input costs. The results of the study highlighted a significant reduction of total costs obtained with both minimum (−16%) and no-tillage (−19%) compared to conventional tillage.

Keywords: conservation agriculture; minimum tillage; no-tillage

---

1. Introduction

Conservation agriculture is a farming system that aims at reducing soil erosion due to intense rainfall and wind phenomena, by promoting the maintenance of a permanent soil cover, minimum soil disturbance, and diversification of plant species. It enhances biodiversity and natural biological processes above and below the ground surface, which contribute to increased water and nutrient use efficiency and to improved and sustained crop production [1].

The practice of conservation agriculture can be summarized by three pillars [2–5]:

(1) reduced soil disturbance by minimizing the mechanized operations and by avoiding inversion tillage (i.e., minimum or no-tillage);
(2) permanent organic cover of soil by crop residues and/or by cover crops between one main crop cycle and the next;
(3) crop rotation and diversification of plant species through varied crop cultivation sequences and/or associations involving at least three different crops.

Major benefits of conservation agriculture practice are also linked to maintaining soil fertility by reducing loss of organic matter and improving structure [6], as well as lowering the release of CO_2 in the atmosphere by enabling the accumulation of carbon in undisturbed soil with carbon sink effect [7–10], and thanks to less use of fossil fuels during tillage [11,12].

Conservation agriculture practice is often associated with the adoption of cover crops, providing specific agronomic advantages, such as improving some physic-chemical properties of the soil and biodiversity [13], in addition to ensuring an adequate protecting cover of the soil until a new crop is grown, with additional decompaction effects and help in controlling soil-borne diseases [14,15].

Significant economic benefits for the farm are also expected from lower production cost, thanks to the reduction of the intensity of mechanized operations with savings of fuel and labor. Indeed, recent studies showed that in specific conditions, the adoption of conservation agriculture can reduce the mechanization costs up to more than 50% in the case of no-tillage farming of corn, and by even more than 75% for common wheat, due to the reduced fuel consumption and the contextual decrease of the work time which results in labor cost savings [6,8,16,17].

However, these savings do no always translate into a greater margin for the farmer, since this depends on the obtained production yield. Contrasting conclusions about the effects of conservation agriculture can be found in the scientific literature. Primarily, results on crop production seem to vary depending on the considered crop, soil, and climate [18]. For example, a multi-year research conducted in the United Kingdom on corn demonstrated that the adoption of minimum tillage techniques increased the gross margin about 6.6 % by reducing the production costs and keeping the yield unchanged [19]. Benefits in crop yield are more evident for cultivation in non-irrigated areas and/or in semi-arid conditions. In these cases, the adoption of conservative practices was associated with increasing soil water holding capacity, leading to higher crop production compared to conventional techniques [20,21].

In contrast, a multi-year research carried out in Italy showed that some crops (corn and wheat, in particular) exhibited a dramatic yield reduction of about 20% after the adoption of minimum tillage and no-tillage [8,22], while for other crops (e.g., soybean) no significant differences were found in comparison with conventional practices [22]. The authors related the reduction to a difficult weed control, and to soil compaction generated by mechanized operations carried out with non-optimal soil moisture conditions.

Concerning paddy rice cultivation, previous research showed that the adoption of minimum tillage did not affect crop yield compared to conventional tillage [8,23,24], while with no-tillage a yield decrease between 10% to 20% was observed [8,23–25]. Again, this was related to difficulty in controlling weeds (in particular *Echinochloa crus-galli* and *Oryza sativa* (L.) var. sylvatica), because of their capability to germinate and grow for extended periods of time in anaerobic conditions.

Nevertheless, given the inherent agronomic and environmental advantages, conservation agriculture was evaluated to be eligible for public subsidies in order to compensate farmers for the possible reduction in production [6,8,22,26].

The scientific literature lacks studies related to mechanization cost analysis for conservation agriculture practice, and typically they are limited to energy and labor costs. For this, the goal of the present study was to experimentally evaluate and analyze the details of total mechanization cost, including ownership and operating costs, of conventional tillage, minimum tillage, and no-tillage techniques applied to paddy rice cultivation in a typical rice farm in Northern Italy, located in a major rice producing district of Europe.

2. Materials and Methods

A field study was conducted in a paddy rice farm located in Pavia province (45° N, 9° E), Italy, one of the main areas of rice production in Europe, characterized by a loam soil (sand: 11%, silt: 52%, clay: 11%). Particle-size classes were determined by dispersing soil samples with Na hexametaphosphate and subsequently by applying the pipette method according to Italian standard methods for chemical analyses of soil [27]. The main soil characteristics of the field were: pH = 7.1, total nitrogen = 2.08 g/kg, phosphorus = 52.6 ppm P_2O_5, potassium = 173.7 ppm K_2O, organic matter content = 3.44%, CEC (Cation Exchange Capacity) = 17.9 meq/100 g, exchangeable calcium = 3124.3 ppm Ca^{2+}, exchangeable magnesium = 208.3 ppm Mg^{2+}, and exchangeable sodium = 40.3 ppm Na^{2+}.

In order to analyze the total mechanization cost associated with conventional, minimum, and no-tillage agronomic practices, three adjacent plots with the same size and shape were obtained from a whole field (Figure 1).

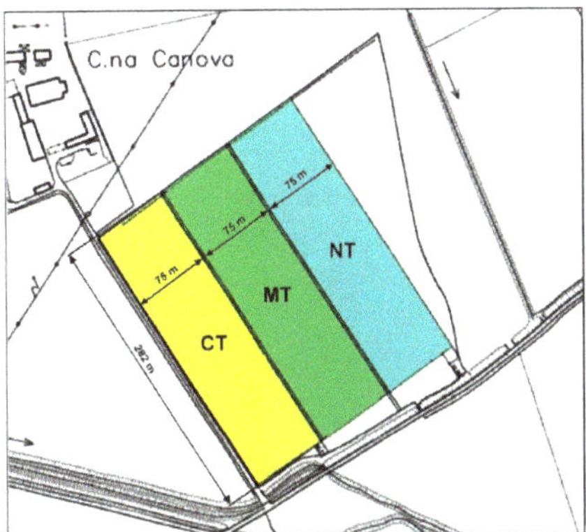

Figure 1. The three experimental plots used for the field study (CT: conventional tillage, MT: minimum tillage, NT: no-tillage).

The three plots were each 2.11 ha in size, and were delimited by means of a topographic survey, and traced with centimeter accuracy by a tractor equipped with a rotary ditcher and an RTK (Real Time Kinematic) automatic guidance system. In the previous three years, the field was cultivated according to minimum tillage practice (non-inversion tillage with working depth of 10 cm) and was regularly sown with cover crops between two main crops. After the last harvest of soybean (October 2017), the field was sown with a cover crop mixture of vetch (*Vicia sativa* L.) and rapeseed (*Brassica napus* L.).

For the experimental tests, rice (*Oryza Sativa* L.) ssp japonica, cultivar *Caravaggio* was sown. This commercial cultivar has a 147-day nominal growth cycle for the farm area, a significant resistance to main fungal diseases (e.g., *Magnaporthe oryzae*), and high yield potential.

The three plots, named CT (conventional tillage), MT (minimum tillage), and NT (no-tillage), respectively, underwent different soil preparation to seeding. The other crop operations—fertilization (3 distributions), pre-emergence (1 application) and post-emergence (1 application) herbicide treatments, fungicide treatments (2 applications), and harvesting—were exactly the same (same dates, same amount, and dose).

Finally, in order to compare fuel consumption and work time associated with the agronomical practices, the same tractors were used in three plots (tractor 1: 144 kW rated engine power; tractor 2: 97 kW rated engine power; both four-wheel drive (4WD)), as it was the same combine harvester used for the harvest.

The soil preparation in the three plots according to the three different practices is detailed as follows.

2.1. Conventional Tillage (CT)

The CT practice started with ploughing with a 4-share reversible plough, coupled to tractor 1. The plough working depth was set at 27 cm, enough for the complete burying of the autumn cover crop. Subsequently, a pre-sowing fertilization with urea (N 46) was carried out by using a centrifugal fertilizer spreader, coupled to tractor 2, followed by one harrowing pass with a working depth of 10 cm to smoot the plot surface. After the secondary tillage, the soil was in ideal conditions to apply the false seedbed technique (i.e., a seedbed preparation earlier than the real sowing date with the purpose of allowing time for germination and subsequent destruction of weeds), ahead of establishing a crop in the real seedbed [16]. Once weeds emerged, a non-selective herbicide (glyphosate) was applied. The formation of a surface crust required a second harrowing conducted by a tine harrow with 3 cm of working depth, in order to create an optimal soil condition for seeding. Sowing was carried out using a pneumatic seed drill

2.2. Minimum Tillage (MT)

The MT practice started with the pre-sowing fertilization of the plot with urea by a centrifugal fertilizer spreader coupled to tractor 2. The tillage operations were limited to disk harrowing with a working depth of 10 cm. Cover crop was properly destroyed, even if with the formation of coarse soil clod that required a second disk harrowing at the same depth but conducted in transverse direction compared to the previous one; this obtained an optimal seedbed both for soil refinement and residue burying. Once tillage operations were concluded, the seeding was delayed in order to allow weeds germination (in particular *Oryza sativa* (L.) var. sylvatica) and subsequent control by a non-selective herbicide treatment. Sowing was carried out using a combined seed drill.

2.3. No-Tillage (NT)

The NT practice began with the pre-sowing fertilization of the plot with urea by a centrifugal fertilizer spreader, followed by a chemical termination of the cover crop with glyphosate. Thereafter, in order to facilitate the sowing, the cover crop residues were lodged in the seeding direction by using a Cambridge roller. The seeding was carried out using a sod-seeder with a preset load of 2.8 kN per planter unit. The seeding rate was the same as the one adopted in CT and MT (200 kg/ha).

2.4. Production Factors

Table 1 shows the sequence of the mechanized operations adopted in the three plots, together with the amount of production input used (seeds, fertilizers, pesticides).

2.5. Field Capacity

The work time of every mechanized operation conducted was recorded by a single frequency GPS (Global Positioning System) receiver (ArvaPc, Arvatec Srl, Milan, Italy) installed on the two tractors used for field operations. The GPS recorded, with a frequency acquisition of 1 Hz, the date, time, and position of the tractor. The generated log file was in NMEA 0183 format. The recorded data were analyzed to compute for every conducted operation the effective field capacity (C_a, ha/h); that is the actual rate of land processed per time unit [28], calculated as follows:

$$C_a = \frac{A}{wt} \tag{1}$$

where, A = area processed by the equipment (ha) (2.11 ha in the case of the operation carried out on the single plot, 6.33 ha regarding operations carried out at the same time on the three plots); wt = total work time measured by the GPS receiver (h), it includes the actual operating time, turnings time, and filling time necessary to refill seed hoppers, fertilizer hopper, and sprayer's tank.

By adding the work time measured for every mechanized operation, the total work time (h) necessary to operate CT, MT, and NT practices in the three plots was computed.

Table 2 shows the main technical parameters of the equipment used and the couplings between tractors and the specific operating machines. Furthermore, Table 3 reports the rated power Pn (kW) for the two 4WD tractors and the working width Dr (m) for the operating machines.

Table 1. Sequence of the mechanized operations carried out on the three experimental plots.

Conventional Tillage (CT)			Minimum Tillage (MT)			No-Tillage (NT)		
Operation	Production Factor	Rate (kg/ha)	Operation	Production Factor	Rate (kg/ha)	Operation	Production Factor	Rate (kg/ha)
Ploughing	-	-	1st Fertilization	Urea	75	1st Fertilization	Urea	75
1st Fertilization	Urea	75	1st Disc Harrowing	-	-	Herbicide Treatment	Glyphosate	2
Disc Harrowing	-	-	2nd Disc Harrowing	-	-	Rolling	-	-
Herbicide Treatment	Glyphosate	2	Herbicide Treatment	Glyphosate	2	Sod Seeding	Caravaggio cv. Seed	200
Tine Harrowing	-	-	Seeding	Caravaggio cv. Seed	200			
Seeding	Caravaggio cv. Seed	200						
			Pre-emergence Herbicide*	Pendimetalin + Clomazone	1.8 + 0.3			
			2nd Fertilization*	Slow-release Nitrogen	105			
			Post-emergence Herbicide*	Penoxsulam + Terbuthylazine + Bromoxinil	2 + 0.15 + 1.2			
			1st Fungicide Treatment*	Tricyclazole	0.3			
			3rd Fertilization*	19-0-22 Complex	180			
			2nd Fungicide Treatment*	Tricyclazole + Azoxystrobin	0.3 + 0.75			
			Harvesting*	-	-			

*: common to all the three agronomic practices.

Table 2. Main technical parameters of the equipment and coupled tractors used in the three plots of the study.

Operation	Conventional Tillage (CT)			Minimum Tillage (MT)			No-Tillage (NT)		
	4WD Tractor	Operating Machine		4WD Tractor	Operating Machine		4WD Tractor	Operating Machine	
	Pn (kW)	Typology	Dr (m)	Pn (kW)	Typology	Dr (m)	Pn (kW)	Typology	Dr (m)
Ploughing	144	4-share Plough	1.7	n.c.	n.c.	n.c.	n.c.	n.c.	n.c.
Disc Harrowing	n.c.	n.c.	n.c.	144	Disk Harrow	5.0	n.c.	n.c.	n.c.
Harrowing	144	Rotary Harrow	5.0	n.c.	n.c.	n.c.	n.c.	n.c.	n.c.
Fertilization	97	Fertilizer Spreader	20.0	97	Fertilizer Spreader	20.0	97	Fertilizer Spreader	20.0
Pesticides Distribution	97	Sprayer	24.0	97	Sprayer	24.0	97	Sprayer	24.0
Breaking Surface Crust	144	Tine Harrow	5.0	n.c.	n.c.	n.c.	n.c.	n.c.	n.c.
Rolling	n.c.	n.c.	n.c.	n.c.	n.c.	n.c.	97	Cambridge Roller	3.0
Seeding	97	Pneumatic Seed Drill	4.5	97	Combined Seed Drill	4.5	97	No-till Drill	3.0

Abbreviations: n.c.: mechanized operation not carried out in the specific agronomic practice; Dr: working width of the various operating machines; 4WD: four-wheel drive.

2.6. Mechanization Costs Calculation

In order to evaluate the possible profitability of MT and NT in comparison to CT, the total costs associated with the use of each typology of equipment were computed applying the methodology using the ASABE EP 496.3 methodology [28]. This is a reference method for accounting agricultural machinery costs by evaluating their annual ownership costs (€/year) and operating costs (€/h) [29,30]. Ownership is independent of machine use, while operating costs are proportional to the utilization of the machine. Total machine costs are the sum of the ownership and operating costs [28]. In particular, ownership costs include equipment depreciation, interest on the investment, taxes, insurance, and housing of the machine [31].

Depreciation is the reduction in the value of a machine with time and use. It is often the largest single cost of machine ownership and considers the salvage value of the machine at the end of its life.

The cost of ownership includes the interest on the money that is invested in the machine. Typically, a loan is used to purchase the machine; in this case the interest rate is known. If a machine is purchased for cash, the relevant interest rate is the rate that could have been obtained if the money had been invested instead of being used to purchase the machine.

Taxes include sales tax assessed on the purchase price of a machine and property tax assessed on the remaining value in any given year. Insurance is usually related to the civil liability in case of an accident. The cost for housing takes into account the investment for the shelter to recover the agricultural machine. The annual cost of shelter is considered to be constant over the life of the machine.

The ownership costs C_o (€/yr) were calculated through the following equation [28]:

$$C_o = P \times \frac{1 - (1 - t_d)^{L+1}}{L} + \frac{1 + (1 - t_d)^{L+1}}{2} \times i + K_2 \tag{2}$$

where, P = purchase price of the machine (€); t_d = depreciation rate of machine (%); L = machine life (yr); i = annual interest rate (%); K_2 = ownership cost factor for taxes, housing, and insurance (usually 1.5 % of P)

Operating costs are the costs associated with use of a machine and include the costs of fuel and oil, repair and maintenance, and labor.

The cost of fuel for the tractor/combine involved was calculated by measuring the actual fuel consumption during each plot operation (see Equation (6)), multiplied by the market price of fuel.

The cost of lubricant oil was calculated by multiplying the market price of oil by the hourly oil consumption (Q_i, kg/h) calculated by the following equation [32]:

$$Q_i = \rho_{oil} \times (0.000239 \times P_r + 0.00989) \tag{3}$$

where, ρ_{oil} = lubricant oil density (0.880 kg/ dm^3); Pr = rated engine power (kW).

Costs for repairs and maintenance are highly variable, depending on the care provided by the farmer. Repair and maintenance cost (Crm, €/h) tend to increase with the size, complexity, and the working hours of the machine [33]:

$$C_{rm} = P \times FR \times \frac{(L \times H_a)^{RF2-1}}{(Sl)^{RF2}} \tag{4}$$

where, P = purchase price of the machine (€); FR = repair and maintenance factor (% of P); L = machine life (yr); H_a = yearly working hours of the specific machine (h/yr); $RF2$ = repair and maintenance factor; Sl = estimated life of the machine (h).

All the parameters of Equations (2) and (4) are listed in Table 3.

Equation (4) (ASABE EP 496.3 [28] modified by Lazzari and Mazzetto [34]) provides the hourly repair and maintenance cost as a function of the yearly working hours of the specific machine.

Ownership, operating, and total machine costs can be calculated on an hourly, or per-ha basis. Total per-ha cost (C$_{tot}$, €/ha·yr) is calculated by dividing the total annual cost of the area covered by the machine during the year, or by the area involved in a particular mechanized activity:

$$C_{tot} = \frac{C_o + (C_{fo} + C_{rm} + C_l) \times H_a}{A} \tag{5}$$

where, C_o = ownership costs (€/yr); C_{fo} = costs for fuel and lubricant oil (€/h); C_{rm} = repair and maintenance costs (€/h); C_l = labor cost (€/h); H_a = yearly working hours of the specific machine (h/yr); A = considered area (ha).

Table 3 lists the economic parameters used for applying the ASABE EP 496.3 methodology [28] for every equipment.

After each operation, the volume of diesel consumed was measured by refilling the fuel tank of the tractor/harvester by using a graduated transparent container, and per-ha fuel consumption (kg/ha) was computed as:

$$\text{Fuel consumption} = \rho_{diesel} \times \frac{x}{A} \tag{6}$$

where, ρ_{diesel} = diesel density (0.835 kg/dm^3); x = volume of diesel consumed for each operation (dm^3); A = area processed by the equipment (2.11 ha in the case of the operation carried out on the single plot, 6.33 ha for the operations carried out at the same time on the three plots).

Table 3. Economic parameters used for applying the ASABE EP 496.3 methodology [28] for every considered equipment.

Agricultural Machine	Purchase Price[*] (€)	Depreciation Rate[**] (%)	Machine Life[**] (yr)	Estimated Life[**] (h)	Annual Interest Rate[*] (%)	FR[**] (%)	RF2[**] (-)	Labor Cost[*] (€/h)
Tractor 1	118,800	12.5	12	12000	3.5	80	2.0	20
Tractor 2	58,000							
4-share Plough	18,800	18	12	2000	3.5	100	1.8	
Disk Harrow	30,000	18	12	2000	3.5	60	1.7	
Rotary Harrow	19,800	19.5	10	2000	3.5	80	2.2	
Fertilizer Spreader	6100	21	8	1500	3.5	70	1.3	
Sprayer	40,000	25.5	6	2000	3.5	60	1.3	
Tine Harrow	18,000	19.5	10	2000	3.5	70	1.4	
Cambridge Roller	6000	19.5	12	2000	3.5	70	1.3	
Pneumatic Seed Drill	45,600	21	8	1500	3.5	75	2.1	
Combined Seed Drill	45,600	21	8	1500	3.5	75	2.1	
No-till Drill	32,000	21	8	1500	3.5	75	2.1	

[*] Typical current values for Italian market. The labor cost is related to the tractor driver only. [**] According to [35].

In order to compute the costs related to diesel and lubricant oil consumption associated to each operation, a price of 1 €/kg for diesel and 3.5 €/kg for lubricant oil was considered [32].

The results obtained for the three plots were then scaled-up to a paddy rice farm area of 75 ha. This farm size was chosen because it is typical for the producing area considered in the study, as well as because the field capacity of the machines considered would accomplish the sequence of operations in the available time for field work, without the need of additional units of equipment.

The total costs per ha was hence obtained by summing the cost of the production factors used (seed, fertilizers, agro-chemicals) and the mechanization costs (including labor cost), calculated for each considered tillage practice. The harvest of paddy rice was made by a combine contractor at a cost of 250 €/ha

3. Results and Discussion

Table 4 shows the dates on which the mechanized operations were carried out in the three experimental plots, and the related effective field capacity (ha/h) calculated from the GPS data recorded during the field activities. As expected, the lowest field capacity was found for conventional tillage (ploughing and harrowing, with an effective field capacity of 0.76 ha/h and 1.5 ha/h, respectively), and for seeding (from 1.2 to 1.7 ha/h), due to the small working width of the machines. On the contrary, the highest field effective capacity was found for operations conducted with large working width (i.e., fertilizations 9.3–10.5 ha/h and protection treatments 8–10.5 ha/h) for all the CT, MT, and NT plots.

Table 4. Measured effective field capacity (C_a, ha/h) of the operations carried out in the three experimental plots. All operations were carried out in 2018.

Conventional Tillage (CT)			Minimum Tillage (MT)			No-Tillage (NT)		
Date	Operation	C_a (ha/h)	Date	Operation	C_a (ha/h)	Date	Operation	C_a (ha/h)
23 April	Ploughing	0.76	n.c.	n.c.	n.c.	n.c.	n.c.	n.c.
24 April	Harrowing	1.5	24 April	1st Fertilization	9.3	n.c.	n.c.	n.c.
24 April	1st Fertilization	9.3	25 April	1st Disc Harrowing	2.7	24 April	1st Fertilization	9.3
15 May	Herbicide Treatment	8	25 April	2nd Disc Harrowing	2.1	15 May	Herbicide Treatment	8
16 May	Breaking Surface Crust	3.6	15 May	Herbicide Treatment	8	17th May	Rolling	3.1
17 May	Seeding	1.7	17 May	Seeding	1.8	17 May	Sod Seeding	1.2
17 May	Pre-emerge Herbicide Treatment	13	17 May	Pre-emerge Herbicide Treatment	13	17 May	Pre-emerge Herbicide Treatment	13
20 June	2nd Fertilization	10.5	20 June	2nd Fertilization	10.5	20 June	2nd Fertilization	10.5
21 June	Post-emerge Herbicide Treatment	13	21 June	Post-emerge Herbicide Treatment	13	21 June	Post-emerge Herbicide Treatment	13
14 July	1st Fungicide Treatment	9.9	14 July	1st Fungicide Treatment	9.9	14 July	1st Fungicide Treatment	9.9
20 July	3rd Fertilization	9.3	20 July	3rd Fertilization	9.3	20 July	3rd Fertilization	9.3
27 July	2nd Fungicide Treatment	9.9	27 July	2nd Fungicide Treatment	9.9	27 July	2nd Fungicide Treatment	9.9
17 October	Harvesting	1.2	17 October	Harvesting	1.2	17 October	Harvesting	1.2

Abbreviation: n.c.: mechanized operation not carried out in the specific agronomic practice.

Table 5 shows the fuel consumption (kg/ha and kg/h of diesel) for every operation. Again, the highest fuel consumption was found for ploughing (34.1 kg/ha of diesel) and rotary harrowing (18.9 kg/ha), both used only in CT practice. Note that ploughing, rotary harrowing, disc harrowing, and the tine surface harrowing were carried out by the 144-kW tractor 1, whilst for the other activities the 97-kW tractor 2 was used. The fuel consumption related to the seeding was the same for CT and MT (6.9 kg/ha), while it was lower for NT (5.7 kg/ha), due to the typology of the seed drill used. Finally, fuel consumption for paddy rice harvesting was 17.3 kg/ha of diesel.

Table 5. Fuel consumption measured during the field operations.

Conventional Tillage (CT)			Minimum Tillage (MT)			No-Tillage (NT)		
Operation	Fuel Cons. (kg/ha)	Fuel Cons. (kg/h)	Operation	Fuel Cons. (kg/ha)	Fuel Cons. (kg/h)	Operation	Fuel Cons. (kg/ha)	Fuel Cons. (kg/h)
Ploughing	34.1	25.9	n.c.	n.c.	n.c.	n.c.	n.c.	n.c.
Harrowing	18.9	28.4	1st Fertilization	0.9	8.4	n.c.	n.c.	n.c.
1st Fertilization	0.9	8.4	1st Disc Harrowing	6.5	17.6	1st Fertilization	0.9	8.4
Herbicide Treatment	1.3	11.4	2nd Disc Harrowing	7	14.7	Herbicide Treatment	1.3	11.4
Breaking Surface Crust	4.4	15.8	Herbicide Treatment	1.3	11.4	Rolling	2.6	8.0
Seeding	6.9	11.7	Seeding	6.9	12.4	Sod Seeding	5.7	7.1
Pre-emergence Herbicide treatment	1.3	16.9	Pre-emergence Herbicide Treatment	1.3	16.9	Pre-emergence Herbicide Treatment	1.3	16.9
2nd Fertilization	0.9	9.5	2nd Fertilization	0.9	9.5	2nd Fertilization	0.9	9.5
Post-emergence Herbicide Treatment	1.3	16.9	Post-emergence Herbicide Treatment	1.3	16.9	Post-emergence Herbicide Treatment	1.3	16.9
1st Fungicide Treatment	1.3	12.9	1st Fungicide Treatment	1.3	12.9	1st Fungicide Treatment	1.3	12.9
3rd Fertilization	0.9	8.4	3rd Fertilization	0.9	8.4	3rd Fertilization	0.9	8.4
2nd Fungicide Treatment	1.3	12.9	2nd Fungicide Treatment	1.3	12.9	2nd Fungicide Treatment	1.3	12.9
Harvesting	17.3	20.8	Harvesting	17.3	20.8	Harvesting	17.3	20.8
Total	90.8	-	Total	46.9	-	Total	34.1	-

n.c.: mechanized operation not carried out in the specific agronomic practice.

Overall, the total fuel consumption for the three agronomic practices was 90.8 kg/ha for CT, 46.9 kg/ha for MT, and 34.1 kg/ha for NT, corresponding to fuel savings of 48% and 63% for MT and NT, respectively, compared to CT.

These findings are quite in accordance with those obtained by Rognoni et al. [8] for wheat cultivation in Italy, with fuel savings of 42% for MT and of 75% for NT, compared to CT. Similarly, for corn, they found 57% (MT) and 61% (NT) savings compared to CT. Studying wheat cultivation in the United Kingdom, Morris et al. [35] obtained fuel savings of 32% for MT and of 77% for NT, compared to CT, but only considering tillage without accounting for the consumption associated with other mechanized operations (fertilizing, pesticides distributions, harvesting).

By scaling up the experimental results obtained in the three plots on a rice farm area of 75 ha, the working hours required by paddy rice cultivation with CT, MT, and NT practices are shown in Table 6. In overall, the working time for CT was 335.4 h, MT was 227.0 h, and NT was 208.5 h, with work savings for MT and NT of 32% and 38%, respectively, compared to CT.

Table 6. Computed work times for 75 ha paddy farm with the three considered agronomic practices.

Conventional Tillage (CT)		Minimum Tillage (MT)		No-Tillage (NT)	
Operation	Work Time Per 75 ha Area (h)	Operation	Work Time Per 75 ha Area (h)	Operation	Work Time Per 75 ha Area (h)
Ploughing	98.7	n.c.	n.c.	n.c.	n.c.
Harrowing	50.0	1st Fertilization	8.1	n.c.	n.c.
1st Fertilization	8.1	1st Disc Harrowing	27.8	1st Fertilization	8.1
Herbicide Treatment	9.4	2nd Disc Harrowing	35.7	Herbicide Treatment	9.4
Breaking Surface Crust	20.8	Herbicide Treatment	9.4	Rolling	24.2
Seeding	44.1	Seeding	41.7	Sod Seeding	62.5
Pre-emergence Herbicide Treatment	5.8	Pre-emergence Herbicide Treatment	5.8	Pre-emergence Herbicide Treatment	5.8
2nd Fertilization	7.1	2nd Fertilization	7.1	2nd Fertilization	7.1
Post-emergence Herbicide Treatment	5.8	Post-emergence Herbicide Treatment	5.8	Post-emergence Herbicide Treatment	5.8
1st Fungicide Treatment	7.6	1st Fungicide Treatment	7.6	1st Fungicide Treatment	7.6
3rd Fertilization	8.1	3rd Fertilization	8.1	3rd Fertilization	8.1
2nd Fungicide Treatment	7.6	2nd Fungicide Treatment	7.6	2nd Fungicide Treatment	7.6
Harvesting	62.3	Harvesting	62.3	Harvesting	62.3
Total	335.4	Total	227.0	Total	208.5

n.c.: mechanized operation not carried out in the specific agronomic practice.

The total time necessary to cultivate one hectare was 4.5 h/ha for CT, 3.0 h/ha for MT, and 2.8 h/ha for NT. Considering an hourly labor cost of 20 €/h, it follows that the labor cost per hectare is 72.8 €/ha for CT, 43.9 €/ha for MT, and 38.9 €/ha for NT. Morris et al. [35] found that the total time necessary for tillage operations on one hectare of wheat is 2.5 h/ha for CT, 1 h/ha for MT, and 0.5 h/ha for NT.

In this study, the main factors of the mechanization operating costs (fuel + labor) resulted in 163.6 €/ha for CT, 90.8 €/ha for MT, and 73.0 €/ha for NT, with savings of 46% and 55% of conservative techniques compared to CT; that was in fair agreement with [36].

By considering the total costs of mechanization for the machines used in the study and scaling up to the case of a 75 ha farm size, the differences in costs were relatively less marked (Table 7) than the comparison to the simple sum of diesel and labor costs for the three considered practices. In fact, the total costs of mechanization for a 75 ha paddy rice farm, calculated through the methodology defined in the ASABE EP 496.3 standard [28] (assuming an annual use of 500 h for both tractors) was 604.8 €/ha for CT, 424.8 €/ha for MT, and 382.7 €/ha for NT, with savings of 30% and 37%, respectively. In a study on soybean cultivation in the USA, McIsaac et al. [17] obtained savings of 16% and 27% for MT and NT, respectively, by only accounting tillage operations and without considering the incidence of production factors and other operations.

Considering that the costs per hectare for seed cv. Caravaggio, fertilizer, and herbicide and fungicide resulted 195.0 €/ha, 129.9 €/ha, and 332.2 €/ha respectively, the total costs (mechanization costs, labor cost, cost for seed, fertilizer, and agro-chemicals) related to the three agronomic practices were finally computed (Table 8). De facto, since the cost for the factors of production is the same for CT, MT, and NT, the observed differences were only due to the mechanization costs for tillage, and to the labor requirement necessary to conclude the considered agronomic practices.

Table 7. Total costs of mechanization by referring a paddy area of 75 ha.

Conventional Tillage (CT)			Minimum Tillage (MT)			No-Tillage (NT)		
Operation	Hourly Cost (€/h)	Cost Per Hectare (€/ha)*	Operation	Hourly Cost (€/h)	Cost Per Hectare (€/ha)*	Operation	Hourly Cost (€/h)	Cost Per Hectare (€/ha)*
Ploughing	82.20	108.16	n.c.	n.c.	n.c.	n.c.	n.c.	n.c.
Harrowing	108.42	72.28	n.c.	n.c.	n.c.	n.c.	n.c.	n.c.
Fertilization	63.29	6.52	Fertilization	63.29	6.52	Fertilization	63.29	6.52
Distribution of Agrochemicals	233.68	21.24	Disc Harrowing	120.01	42.50	Distribution of Agrochemical	233.68	21.24
Breaking Surface Crust	145.46	40.41	Distribution of Agrochemicals	233.68	21.24	Rolling	53.29	17.19
Seeding	180.60	106.24	Seeding	188.20	104.56	Sod Seeding	105.30	87.75
Harvesting	-	250.00	Harvesting	-	250.00	Harvesting	-	250.00
Total	-	604.85	Total	-	424.82	Total	-	382.71

Abbreviations: n.c.: mechanized operation not carried out in the specific agronomic practice. * Calculated for every mechanized operation as the ratio between the hourly cost and the effective field capacity C_a.

Table 8. Total costs of paddy rice production according the three agronomic practices on a paddy area of 75 ha, and the savings achievable by MT and NT in comparison with CT.

Costs Per Hectare (€/ha)	Conventional Tillage (CT)	Minimum Tillage (MT)	No-Tillage (NT)
Mechanization Costs	604.85	424.82	382.71
Labor Cost	72.79	43.87	38.94
Seed	195.0	195.0	195.0
Fertilizer	129.90	129.90	129.90
Pesticides	332.2	332.2	332.2
Total	1334.74	1125.79	1078.75
Saving in Comparison with CT (%)	-	16 %	19 %

Finally, considering the total costs to produce paddy rice, including mechanization, labor, seed, fertilizer, and pesticides, the total cost per hectare amounted to 1334.7 €/ha for CT, 1125.8 €/ha for MT, and 1078.7 €/ha for NT, with total savings of 16% and 19%, respectively These findings demonstrate that from the production costs point of view, conservation agriculture can be more sustainable than conventional approaches. It should be recalled, however, that conservation agriculture techniques do not always allow levels of production comparable with those obtained with conventional approaches. In the case of a decrease in yield, despite the reduction of mechanization costs due to the conservation approaches, the economic balance can be uncertain for farmers.

4. Conclusions

This study aimed to contribute to filling the gap in the scientific literature about the technical-economic analysis of conservation agriculture approaches in paddy rice cultivation, compared with conventional practices. With a comparative experiment based on direct field measurements of data about machinery and production factors used, the analysis showed that the adoption of conservative techniques for paddy rice cultivation allowed significant savings on production costs, thanks to reduced work time (47%–61% less) and to lower mechanization costs (42%–58% less) in comparison to conventional practices. Moreover, the reduction found in fuel consumption (48%–63% less) with the associated reduction of emissions can also be related to the direct environmental benefits.

Author Contributions: Conceived and designed the experiments, performed the experiments, collected and analyzed the data, interpreted the results and developed the manuscript, A.C.; conceived and designed the experiments, performed the experiments, collected and analyzed the data, interpreted the results and developed the manuscript, R.O.

Funding: This research received no external funding.

Acknowledgments: The authors would like to acknowledge the Azienda Agricola Sgariboldi (Torrevecchia Pia, Pavia, Italy) for hosting the field tests.

Conflicts of Interest: The authors declare no conflict of interest.

References

1. FAO. What Is Conservation Agriculture? Available online: http://www.fao.org/ag/ca/1a.html (accessed on 30 November 2018).
2. Baker, C.J.; Saxton, K.E.; Ritchie, W.R.; Chamen, W.C.T.; Reicosky, D.C.; Ribero, F.; Justice, S.E.; Hobbs, P.R. *No-Tillage Seeding in Conservation Agriculture*, 2nd ed.; Food and Agriculture Organization of the United Nation (F.A.O.): Rome, Italy, 2014; pp. 1–391.
3. Derpsch, R.; Friedrich, T.; Kassam, A.; Hongwen, L. Current status of adoption of no-till farming in the world and some of its main benefits. *Int. J. Agr. Biol. Eng.* **2010**, *3*, 1–25.
4. Derpsch, R.; Moriya, K. Historical review of no-tillage cultivation of crops. In *Implications of No-Tillage Versus Soil Preparation on Sustainability of Agricultural Production, Proceedings of the 9th Conference of the International-Soil-Conservation-Organisation, Bonn, Germany, 26–30 August 1998*; Blume, H.P., Eger, H., Fleischhauer, E., Hebel, A., Reij, C., Steiner, K.G., Eds.; Catena Verlag: Reiskirchen, Germany, 1998; Volume 31, pp. 1179–1186.
5. Van den Putte, A.; Goversa, G.; Dielsa, J.; Gillijns, K.; Demuzerea, M. Assessing the effect of soil tillage on crop growth: A meta-regression analysis on European crop yields under conservation agriculture. *Eur. J. Agron.* **2010**, *33*, 231–241. [CrossRef]
6. Pisante, M.; Stagnari, F. *Agricoltura blu. La via Italiana Dell'agricoltura Conservativa*; Associazione Italiana per la Gestone Agronomica e Conservativa del Suolo: Ancona, Italy, 2011; pp. 1–50.
7. Lindstrom, M.J.; Schumacher, T.E.; Cogo, N.P.; Blecha, M.L. Tillage effects on water runoff and soil erosion after sod. *J. Soil Water Conserv.* **1998**, *53*, 59–63.
8. Rognoni, G.L.; Finzi, A.; Cavalchini, A.G. *Agricoltura Conservativa: Definizioni, Aspetti Tecnici, Esperienze e Opportunità per L'agricoltura Della Provincia di Pavia*; Provincia di Pavia: Pavia, Italy, 2014; pp. 1–51.
9. Sheehy, J.; Regina, K.; Alakukku, L.; Six, J. Impact of no-till and reduced tillage on aggregation and aggregate-associated carbon in Northern European agroecosystems. *Soil Tillage Res.* **2015**, *150*, 107–113. [CrossRef]
10. Soane, B.D.; Ball, B.C.; Arvidsson, J.; Basch, G.; Moreno, F.; Roger-Estrade, J. No till in northern western and south Western Europe. A review of problems and opportunites for crop production and the environment. *Soil Tillage Res.* **2012**, *118*, 66–87. [CrossRef]
11. Tebrügge, F.; Böhrnsen, A. Crop yields and economic aspects of no-tillage compared to plough tillage: Results of long-term soil tillage field experiments in Germany. In Proceedings of the EC Workshop-IV Experience with the Applicability of No-Tillage Crop Production in the West-European Countries, Giessen, Boigneville, France, 12–14 May 1997; Tebrügge, F., Böhrnsen, A., Eds.; Dr Flesh: Langgöns, Germany, 1997; pp. 25–43.
12. Alluvione, F.; Moretti, B.; Sacco, D.; Grignani, C. EUE (energy use efficiency) of cropping systems for a sustainable agriculture. *Energy* **2011**, *36*, 4468–4481. [CrossRef]
13. De Torres, R.-R.; Carbonell-Bojollo, R.; Alcántara-Braña, C.; Rodríguez-Lizana, A.; Ordóñez-Fernández, R. Carbon sequestration potential of residues of different types of cover crops in olive groves under mediterranean climate. *Span. J. Agric. Res.* **2012**, *10*, 649–661. [CrossRef]
14. Rodríguez-Lizana, A.; Repullo-Ruibérriz de Torres, M.A.; Carbonell-Bojollo, R.; Alcántara, C.; Ordóñez-Fernández, R. Brachypodium distachyon, Sinapis alba, and controlled spontaneous vegetation as groundcovers: Soil protection and modeling decomposition. *Agric. Ecosyst. Environ.* **2018**, *265*, 62–72. [CrossRef]
15. Kadziene, G.; Suproniene, S.; Auskalniene, O.; Pranaitiene, S.; Svegzda, P.; Versuliene, A.; Ceseviciene, J.; Janusauskaite, D.; Feiza, V. Tillage and cover crop influence on weed pressure and Fusarium infection in spring cereals. *Crop Prot.* **2020**, *127*, 104966. [CrossRef]
16. Jensen, P.K. Use of integrated weed management tools in crop rotations with grass seed production. *Acta Agric. Scand. Sect. B Soil Plant Sci.* **2019**, *69*, 209–218. [CrossRef]
17. McIsaac, G.F.; Siemens, J.C.; Hummel, J.W.; Tyrrell, A.T. Economic comparison of six corn and soybean tillage systems, two soybean rows spacing on three farm sizes. *Appl. Eng. Agric.* **1990**, *6*, 557–564. [CrossRef]

18. Sharma, P.; Abrol, V.; Sharma, R.K. Impact of tillage and mulch management on economics, energy requirement and crop performance in maize-wheat rotation in rainfed subhumid inceptisols, India. *Eur. J. Agron.* **2011**, *34*, 46–51. [CrossRef]

19. Chetan, F.; Rusu, T.; Chetan, C.; Moraru, P.I. Influence of soil tillage upon weeds, production and economic efficiency of corn crop. *AgroLife Sci. J.* **2016**, *5*, 36–43.

20. Rusinamhodzi, L.; Corbeels, M.; van Wijk, M.T.; Rufino, M.C.; Nyamangara, J.; Giller, K.E. A meta-analysis of long-term effects of conservation agricolture on maize grain yield under rain-fed condition. *Agron. Sustain. Dev.* **2011**, *31*, 657–673. [CrossRef]

21. Pittelkow, C.M.; Liang, X.; Linquist, B.A.; van Groenigen, K.J.; Lee, J.; Lundy, M.E.; van Gestel, N.; Six, J.; Venterea, R.T.; van Kessel, C. Productivity limits and potentials of the principles of conservation agriculture. *Nature* **2015**, *517*, 365–368. [CrossRef] [PubMed]

22. Bolzonella, C.; Boatto, V. All'agricoltura conservativa serve il sostegno pubblico. *L'informatore Agrar.* **2015**, *15*, 54–58.

23. Ente Nazionale Risi. XLVI relazione annuale. *Il Risicoltore* **2013**, *2*, 1–100.

24. Ente Nazionale Risi. XLVIII relazione annuale. *Il Risicoltore* **2015**, *2*, 1–100.

25. Moretti, B.; Celi, L.; Vitali, A.; Lerda, C.; Sacchi, G.; Maghrebi, M.; Abruzzese, A.; Beltarre, G.; Romani, M.; Sacco, D. Ottimizzare le produzioni risicole con l'agricoltura conservativa. *L'Informatore Agrar.* **2019**, *24–25*, 31–34.

26. Kassam, A.; Friedrich, T.; Derpsch, R.; Kienzle, J. Overview of the worldwide spread of conservation agriculture. *FACTS Rep.* **2015**, *8*, 1–11.

27. Italian Ministry for Agriculture and Forestry Policy. *Analytical Methods for Chemical Analyses of Soils (In Italian)*; Franco Angeli: Milano, Italy, 2000.

28. ASABE. *ASABE Standard EP496.3 FEB2006 (R2015): Agricultural Machinery Management*; ASABE: St. Joseph, MI, USA, 2015.

29. Tona, E.; Calcante, A.; Oberti, R. The profitability of precision spraying on specialty crops: A technical–economic analysis of protection equipment at increasing technological levels. *Precis. Agric.* **2018**, *19*, 606–629. [CrossRef]

30. Tangorra, F.M.; Calcante, A. Energy consumption and technical-economic analysis of an automatic feeding system for dairy farms: Results from a field test. *J. Agric. Eng.* **2018**, *49*, 228–232. [CrossRef]

31. Srivastava, A.K.; Goering, C.E.; Rohrbach, R.P.; Buckmaster, D.R. *Engineering Principles of Agricultural Machines*; American Society of Agricultural and Biological Engineers: St. Joseph, MI, USA, 2006; pp. 1–588.

32. Calcante, A.; Brambilla, M.; Bisaglia, C.; Oberti, R. Estimating the total lubricant oil consumption rate in agricultural tractors. *Trans. ASABE* **2019**, *62*, 197–204. [CrossRef]

33. Calcante, A.; Fontanini, L.; Mazzetto, F. Repair and maintenance costs of 4WD tractors in Northern Italy. *Trans. ASABE* **2013**, *56*, 355–362. [CrossRef]

34. Lazzari, M.; Mazzetto, F. *Mechanics and Mechanization of Agricultural Process (in Italian: Meccanica e Meccanizzazione dei Processi Produttivi Agricoli)*; Reda: Torino, Italy, 2016; pp. 71–72.

35. ASABE. *ASABE Standard D497.6 JUN2009 (R2015): Agricultural Machinery Management Data*; ASABE: St. Joseph, MI, USA, 2015.

36. Morris, N.L.; Miller, P.C.H.; Orson, J.H.; Froud-Williams, R.J. The adoption of non-inversion tillage systems in the United Kingdom and the agronomic impact on soil, crops and the environment—A review. *Soil Tillage Res.* **2010**, *108*, 1–15. [CrossRef]

MDPI

St. Alban-Anlage 66

4052 Basel

Switzerland

Tel. +41 61 683 77 34

Fax +41 61 302 89 18

www.mdpi.com

Agronomy Editorial Office

E-mail: agronomy@mdpi.com

www.mdpi.com/journal/agronomy